DATE DUE

Dear Professor Einstein

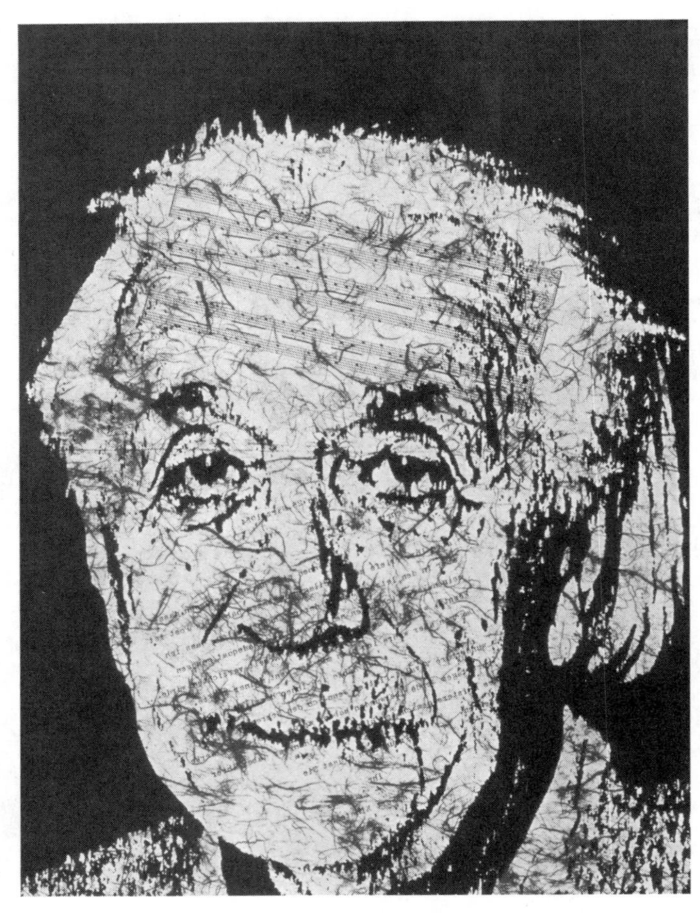

Silkscreen collage overlayed on a photo of Einstein taken by a child in Princeton. (Courtesy of Andor Orand)

Dear Professor Einstein

Foreword by
Evelyn Einstein

Albert Einstein's
Letters to
and from Children

Edited by Alice Calaprice

With an essay by Robert Schulmann

59 John Glenn Drive
Amherst, New York 14228-2197

Published 2002 by Prometheus Books

Inquiries should be addressed to
Prometheus Books
59 John Glenn Drive
Amherst, New York 14228–2197
VOICE: 716–691–0133, ext. 207
FAX: 716–564–2711
WWW.PROMETHEUSBOOKS.COM

06 05 04 03 02   5 4 3 2 1

Library of Congress Cataloging-in-Publication Data

Einstein, Albert, 1879–1955.
Dear Professor Einstein : Albert Einstein's letters to and from children / edited by Alice Calaprice ; foreword by Evelyn Einstein.
p. cm.
Indluces index.
ISBN 1–59102–015–8 (cloth : alk. paper)
1. Einstein, Albert, 1879–1955—Correspondence. 2. Children—Correspondence. 3. Physicists—Correspondence. I. Calaprice, Alice. II. Title.

QC16.E5 A4 2002
530'092—dc21
[B]

2002073570

Printed in Canada on acid-free paper

Dedicated to the children of the world

The author will donate a portion
of the author's royalties
of this book to UNICEF

Contents

Foreword

It is a boon as well as a burden to have the Einstein name all of one's life. My first encounters with people would probably have been different if I had introduced myself as Evelyn Smith or Evelyn Jones. The expectations that others had of me seemed to be higher because I was "an Einstein." So, if I did well in school, there was never any special praise, yet the children still thought of me as a teacher's pet. I was always fearful that people would be angry with me if I did not live up to my name.

My grandfather died when I was fourteen, so naturally all contact ended at that time. I wish I could have known him in my adulthood so that we could have shared experiences, ideas, and speculations. Though he never tried to influence my thoughts, I still feel a great kinship with him in terms of our worldview.

I was born just a few months before the United States entered World War II, at a time when my grandfather was occupied with war and humanitarian concerns at his home in Princeton, New Jersey. I have no recollection of our first meeting, but it must have been during the early 1940s, before the end of the war, when I was just a toddler and we were on a visit from North Carolina. I've been told that I went sailing with him on his sailboat *Tinnef,* and that he was impressed that I loved to sail at my young age, and that I wasn't afraid of the water. After the war, my family moved to California, and our contacts with Princeton became less frequent. At that time, it was not so easy as it is today to hop onto an airplane for visits to see one's grandpa for the holidays. And though my grandfather had visited California in the 1930s, he did not leave the East

Coast after he moved to Princeton because he was always so busy. So most of what I knew and heard about him came from a distance, but I never thought of him as anything more than my grandfather.

In 1953 I was sent off to Switzerland to boarding school, and I came back home only once until 1958. Because my parents were from Europe, they wanted me to be raised culturally as a European, though I know it must have been hard for my mother to send me away for so long. She was a warm and kind woman who had been devoted to her children, but my father was more distant and was able to remove himself from our lives more easily. Therefore, aside from the sailing episode when I was a toddler, my memories of seeing my grandfather stem from the times we stopped in Princeton on the way to Europe and on the way back.

My grandfather lived on Mercer Street in a two-story white house with a front porch and columns. A staircase led upstairs to his office. Except when he was busy, I felt welcome in his presence because he made me feel very much at ease. Even when I was only five, he never talked down at me or intimidated me. I was astonished, because in my experience with adults, it was best to stay out of their way. Maybe I was a welcome diversion from other things that were going on in his life. He loved wooden toys and puzzles that came apart and could be put back together again—this is how I remember playing with him.

I loved astronomy as a child. I loved looking out the window at night at the enormous, dark clear sky above me. I first became aware of the wonders of the sky when I was in summer camp in the central valley of California and when we spent time in the mountains. And later, in boarding school, the stars seemed especially beautiful in the alps of Switzerland. I hunted for constellations and let my imagination go wild with thoughts of space travel in the outer reaches of the universe. My grandfather knew about my galactic fascinations as a child and gave me a book I still have today called *The Stars for Sam*. He inscribed it for me with a loving message, appreciating my curiosity about the universe. I maintained my interest in the reaches of outer space throughout my adulthood and even became an ardent *Star Trek* fan.

My grandfather and I continued to correspond

into my teen years until he died. I would tell him about the things that were going on in my life, and he would make comments acknowledging that they were of interest to him. And, like my father before me, I would write and ask him to help me with geometry problems, and he never failed to reply, sometimes telling me I had used a clever approach to solving a problem.

In this collection of letters, therefore, the ones that I appreciate most are those that my grandfather exchanged with the South African girl, Tyfanny, who was spending her high school years in a boarding school and liked to look up at the sky. Tyfanny could easily have been me. Perhaps that's why my grandfather took the time to correspond with her, because she reminded him of me.

This collection is a fine representation of the esteem that children had for my grandfather, and of his willingness to respond to some of them even though he was busier than most people. He respected children and liked their curiosity and fresh approach to life and therefore did not want to ignore them. It is clear to me, however, that, unlike most of today's letters from celebrities, his words are his own. I hope the letters will inspire kids to partake again in the declining art of correspondence so that legacies such as these can be left for future generations.

Evelyn Einstein
January 2002

Preface

In my previous books on Einstein, I published documented sound bites from Einstein's various writings in an attempt to separate the mythological Einstein from the real human being. A characteristic of myths and folklore is that there is no correct or final version—people embellish and personalize what they hear, and then pass along their own interpretations in the retelling. This has often been done with Einstein's words, in that people attach his name to words they think will give credibility to their cause or idea but which he more than likely never said. But it is not a myth that Einstein respected and admired children, even though he sometimes may have sounded harsh in chastising them. The letters and a few other documents pertaining to children are presented here in full (except in two cases of excerpts from long letters to his own sons) so that there is no chance of future embellishment or misrepresentation. Readers who want a more complete picture of the "Man of the Century" can consult the supplemental biographical and bibliographical material included at the back of this book.

Letters from and to children are always appealing, especially so when they are written to someone famous. Albert Einstein is known to have loved children, and many of them must have sensed this affection and therefore felt free to write to him over the years. Obviously, he did not have time to answer them all. Yet some of the letters, questions, or the children themselves must have held special appeal to him, and they are the ones who were lucky enough to receive a reply. The simple tone and nature of this correspondence make them an ideal compilation for readers of all ages.

Some of the approximately sixty children's letters you will read in this little book were written with all the innocence and spelling errors of childhood, while others display a cockiness, an undoubting awareness of Einstein's fame, or some prodding by a parent. Still others were the result of letter-writing projects assigned by schoolteachers to children who appear to have had no real inkling of Einstein's contributions to the world. Yet, most youngsters were aware that for some reason Einstein was famous and deserved respect. Many of the children had a genuine interest in science and were not shy about going right to the top for the answers to their questions. Such letters are timeless, for surely many children still ask the same questions today: "Do scientists pray?" "Is time the fourth dimension?" "What holds the Sun and planets in space?" "What is time? What is the soul? What are the heavens?" "If nobody is around and a tree falls, is there a sound?" and even "Do you consider yourself a genius?" Unfortunately, the children did not get answers to all of these intriguing questions from the busy man. No doubt many of them returned from their mailboxes disappointed after weeks of waiting for a response.

What is evident in the letters is that the children want Einstein to know who *they* are, too. They tell us how old they are, where they live, who their siblings are, what subjects are of interest to them, and that, frankly, they are having difficulties in math. Some try to sound grown up or precocious while others write unselfconsciously like the children they still are.

This collection of letters represents children from all over the world—Japan, South Africa, Holland, Germany, England—but mostly they are from American children during the time when Einstein lived in Princeton, New Jersey. The letters from Einstein reprinted here are, as far as I know, the only letters filed away in the Einstein Archive, though it is likely that handwritten letters were sent off in the days before Einstein had a secretary to type them up and file the copies. I have noted which letters were originally written in German, either by Einstein or by the children, and translated them into English. I have also included two letters written by adults on behalf of children or about children.

I thank Princeton University Press and the Hebrew University of Jerusalem, particularly Adriana Yanai, for giving me permission to publish Einstein's letters to the children and providing a number of photographs; Robert Schulmann, former director of the Einstein Papers Project, for agreeing to write the essay on Einstein's education, which should be of interest to teachers and parents; and Evelyn Einstein for recalling anecdotes of Einstein as grandfather. I am grateful to Linda Regan of Prometheus Books for seeing the possibility of a book in these letters and Peggy Deemer for publishing them between two covers with such interest and care. I thank the various archives I contacted, particularly the Leo Baeck Institute and Viola Voss, for giving me permission to use photographs from their collections, and I think Andor Carius, Gillett Griffen, and Todd Yoder for allowing me to use the ones in their personal collections.

Above all, I hope that readers get as much pleasure out of reading this book as I got in putting it together for them.

Princeton, N.J.
June 2002

A Note to the Children

The letters written to Einstein in this little volume span the years 1928 to 1955, the year of Einstein's death, and they come from many parts of the country and the world. All the children who wrote these letters are by now in middle age and even in old age, and no doubt some are no longer with us. I would have liked to have been able to write to each and every person to see if he or she would have any objection to my use of the letters in this book. Obviously, after so many years, locating all the writers is an impossible task, though I have made an effort to do so through the Internet. Indeed, I found one serendipitously when a relative contacted me. I have therefore, except in a few cases of previously published letters, used only first names or initials to avoid any feelings of invasion of privacy. If anyone recognizes a letter as his or her own, I would be happy to hear from that person and make a confirmation.

I will donate a portion of my royalties gained from this book to the United Nations International Children's Educational Fund (UNICEF).

Chronology of Einstein's Life

1879 March 14, Albert Einstein is born in Ulm, Germany.

1880 Family moves to Munich.

1881 November 18, sister Maja is born.

1884 His father gives him a compass, which makes a great impression on him.

1885 In the fall, enters the Petersschule, a Catholic primary school. Also receives Jewish religious instruction at home and begins violin lessons.

1888 Enters the Luitpold-Gymnasium in Munich.

1889 Begins his interest in physics, mathematics, and philosophy.

1894 Einstein family moves to Italy without Albert, who stays behind to finish school. But he leaves school at the end of the year and joins his family.

1895 Takes the exam two years early to enter the Swiss Federal Institute of Technology (FIT; also known as the Polytechnical Institute, or Poly, and now the ETH), but fails the "general knowledge" part and is not admitted. Enrolls in a high school near Zurich for a year to complete his secondary education.

1896 Gives up his German citizenship and remains stateless for five years. Enters the Polytechnical Institute in October.

1900 Graduates from the Polytechnical Institute. Tells his mother he plans to marry fellow-student Mileva Marić, and she disapproves. Has difficulty finding a job, but sends his first scientific paper to the German physics journal *Annalen der Physik.*

1901 Becomes a Swiss citizen. Continues to look for work and obtains some tutoring jobs. Submits a doctoral dissertation to the University of Zurich. Continues his relationship with Mileva Marić. In December, applies for work at the Swiss patent office in Bern.

1902 Probably in January, daughter Lieserl is born. In June, begins work at the Swiss patent office. In October, his father dies in Milan.

1903 January 6, marries Mileva in Bern, and they continue to live there. Lieserl appears to have been left in the care of Mileva's friends or family members. She may have died of scarlet fever at the age of two.

1904 Son Hans Albert is born May 14.

1905 Einstein's most prolific year of doing revolutionary scientific work. He submits five important papers for publication, including the one on the special theory of relativity.

1906 Receives his doctorate from the University of Zurich, but remains employed at the patent office.

1908 Becomes a lecturer at the University of Zurich.

1909 Accepts a position as professor of theoretical physics at the University of Zurich and resigns his position at the patent office.

1910 Second son, Eduard ("Tete"), is born.

1911 Accepts a position at the German University of Prague.

1912 Becomes reacquainted with his cousin Elsa Löwenthal, who will become a major figure in his life. Accepts an appointment as professor of theoretical physics at the Polytechnical Institute in Zurich.

1913 Resigns from the Poly and accepts a position in Berlin at the University of Berlin and the Kaiser Wilhelm Institute of Physics.

1914 April, arrives in Berlin and starts his new position. Mileva and the children come, but they soon leave to return to Zurich. August, World War I begins.

1915 Writes a paper on the general theory of relativity.

1916 Publishes "The Origins of the General Theory of Relativity" and three papers on quantum theory. Becomes president of the German Physical Society.

1917 Becomes ill with a liver ailment and ulcers, and Elsa takes care of him. Becomes director of the Kaiser Wilhelm Institute of Physics.

1919 Is divorced from Mileva and marries Elsa four months later. Elsa's two daughters live with them. His fame begins as his predictions for the general theory of relativity are confirmed during a solar eclipse.

1920 His mother dies in Berlin. German anti-Semitism becomes noticeable, as does opposition to the theory of relativity. Becomes more and more involved in nonscientific interests, such as politics and Zionism.

1921 Makes his first trip to the United States to lecture at Princeton University and to raise funds on behalf of a Hebrew university in Jerusalem.

1922 October through December, takes a trip to Japan, also visiting other cities in the Far East. While in Shanghai, is notified that he received the Nobel Prize in physics for 1921.

1923 Visits Palestine and Spain.

1925 Travels to South America. Becomes an ardent pacifist and supporter of Gandhi.

1928 Becomes ill with a heart problem and remains weak for a year. Helen Dukas becomes his secretary and remains with him as secretary and housekeeper for the rest of his life.

1930 December, visits New York and Cuba, then stays at Caltech in Pasadena, California, until March of the following year.

1931 Visits Oxford, then vacations in his summer home in Caputh, near Berlin. December, embarks for Pasadena again.

1932 At Caltech from January through March. Returns to Berlin. Later, agrees to accept a position at the Institute for Advanced Study in Princeton, New Jersey. December, visits the United States again.

1933 January, Nazis come to power. Gives up German citizenship and does not return to Germany. Temporarily goes to Belgium and Oxford. September, leaves for the United States with Elsa, secretary Helen Dukas, and an assistant. Elsa's children, Margot and Ilse, remain in Europe with their husbands. Begins professorship at the Institute for Advanced Study.

1934 Ilse dies in Paris at age of thirty-seven. Margot and her husband come to Princeton.

1935 Moves to 112 Mercer Street in Princeton, where he continues to live until his death.

1936 December, Elsa dies after a long illness with heart and kidney disease.

1939 Sister Maja comes to live on Mercer Street. August 2, signs the famous letter to President Franklin D. Roosevelt on the military implications of atomic energy.

1940 Receives U.S. citizenship. Maintains dual U.S. and Swiss citizenship until his death.

1941 December, the United States enters World War II.

1945 World War II ends. Retires officially from the Institute for Advanced Study but keeps an office there until his death.

1946 Einstein advocates a world government, declaring that it is the only way to maintain world peace.

1948 Mileva dies in Zurich. Einstein is told that he has a large aneurysm of the abdominal aorta.

1951 Sister Maja dies in Princeton.

1952 Is offered the presidency of Israel, which he declines.

1955 April 11, agrees to sign a joint manifesto with Bertrand Russell urging all nations to renounce nuclear weapons. April 13, his aneurysm ruptures. April 18, Einstein dies at Princeton Hospital at 1:15 A.M. of a ruptured arteriosclerotic aneurysm of the abdominal aorta. He had opposed any life-prolonging surgery, but the autopsy showed it would not have extended his life anyway. His body is cremated the same day, after unauthorized removal of his brain and eyes, and his ashes are scattered by two friends.

"I Am Merely Curious"

A Short Biography

Over the last century, Albert Einstein's name has come to mean "genius." In his lifetime, Einstein was amused that he was thought of in this way, because he believed he was just more curious than others, seeking answers to questions that most adults would have forgotten to ask once they were past childhood. His worldwide fame began when his theory of relativity was confirmed in 1919 by some astronomers, and when soon afterwards he was awarded the Nobel Prize in physics for 1921. His friendly and grandfatherly face, his wild hair and rumpled clothes, his compassion and concern for others, his love of a simple life, his humility and independent spirit, and, yes, his genius—all of these traits combined to make him an unforgettable figure whom people wanted to get to know. At the same time, the widespread use and enjoyment of the telephone, radio, and movies made it easier to spread his fame. One might say he was the world's first media celebrity.

Albert Einstein was born on March 14, 1879, in Ulm in the southern section of Germany called Württemberg, about eighty-five miles west of the larger city of Munich, to middle-class Jewish parents. It was the year of the Zulu wars in southern Africa and Indian uprisings in the American West, and the infamous Joseph Stalin, who was to become the dictator of the communist Soviet Union, was born then, too. Thomas Edison invented the incandescent lightbulb, George Eastman patented a process to make taking pictures easier, and Frank Woolworth opened his first "dime store."

Until a few years before Albert was born, Württemberg had been an independent state with its own king. Then, in 1871, it joined with other independent German states to become part of a new German empire. The leader of the new nation, Otto von Bismarck, was from the north, from the state of Prussia, whose residents were known for their strict discipline, obedience to their elders and to authority, and deep respect for those of high standing in society. The easygoing Württembergers were soon forced into a way of life that had been unfamiliar to them until now. This new way of doing things also crept into the school system, which became very authoritarian, with strict rules about behavior, punctuality, studying, and following a specific and tightly controlled curriculum. It was a common practice to humiliate poor students in front of their classmates. "To me the worst thing seems to be for a school principally to work with the methods of fear, force, and artificial authority. Such treatment destroys the sound sentiments, the sincerity, and the self-confidence of the pupil," Einstein would say later in life. Einstein's personality was not compatible with this strict and narrow-minded approach to education. So he decided to study, on his own, the subjects that were of interest to him but were not offered in the classroom. He continued to follow a course of independent thinking and decision making later in his life, too. He also disliked the authority and rankings of the military system, in which men would blindly follow a leader no matter what he said or what they personally might believe. Later he would say to himself, "That a man can take pleasure in marching in formation to the strains of a band is enough to make me despise him," and "To punish me for my contempt for authority, Fate has made me an authority myself."

Young Albert did not begin to speak until he was about two and a half years old, when his younger sister, Maja, was born. His parents had thought that he might be mentally retarded, especially since he had also looked strange at birth, with a head that was larger than normal. But their fears proved to be unfounded, and he continued to develop normally. When he was told that he would soon have a baby sister, he had imagined some kind of a new toy that he could play with. When he saw baby Maja, he said, "Yes, it is nice, but where are its wheels?" Later, when he spoke more fluently, he adopted the curious habit of quietly moving his lips and repeating to himself every sentence he uttered. He retained this strange behavior until he was seven years old.

At the age of five, Albert began his education at home with a tutor, and he also started to take violin lessons. Music had an especially calming effect on the young boy, for although he was generally a quiet child, he was prone to temper tantrums, striking out sometimes at his teacher and little sister. He proved to have musical talent and enjoyed music until the end of his life, both as a listener and as a player of the violin and piano. "If I were not a physicist," he would say in an interview in 1929, "I would probably be a musician. I live my daydreams in music. I see my life in terms of music. I get most joy in life out of music." (But later on he would also say variously that he would become a plumber, a salesman, or a lighthouse keeper—obviously he was a man of many interests!) At this time his parents also trained him to be independent, going through busy streets on his own and learning not to expect others to fulfill all of his wishes.

That year, little Albert's father brought home a compass for him. This was a turning point in his young life, for he was struck with awe and wonder that the needle always pointed to the magnetic north, as if moved by an invisible hand. It showed him that there were forces in nature that one could not see, and this magic left a lasting impression on him.

At the age of seven he entered public school at the Petersschule in Munich where his family now lived. There he was considered a rather average student. He always required much time to think before giving his answers, so the teachers were not aware that any special scientific skills were in the making. In fact, they would lose patience with him and whack him over the hands with a stick if he didn't answer right away, a practice that was common in German schools in even more modern times. Strangely, this kind of force was considered to be one of the best ways to teach a child to think quickly.

At this time, young Albert also received the required Catholic religious instruction at school and

The Petersschule in Munich, the primary school Albert attended in the third and fourth grades.
(Albert Einstein Archives, Hebrew University of Jerusalem, Israel)

was introduced to Judaic studies by a distant relative at home. Intrigued by what he read, he became passionate in his religious thinking and actions, wanting above all to please God. This he did without any influence from his parents, who were not particularly religious. Later on, his religious feelings turned more to philosophical thought rather than belief in a personal God who could control people and events. Throughout his life he continued to have a strong sense of obligation, especially toward those who were treated unjustly.

At the age of eight and a half, Albert entered the Luitpold-Gymnasium in Munich. A Gymnasium is the German equivalent of a secondary school with a college-prep curriculum. According to its tradition, this particular school emphasized the humanities and classical languages, so Albert did not receive an outstanding education in mathematics and the natural sciences at this time. But his curiosity about these subjects motivated him to read on his own and to teach himself whatever sparked his interest. Soon he was able to prove mathematical theorems and solve difficult problems, even finding an entirely original proof for the Pythagorean theorem in geometry. He was encouraged in this kind of study by his family and by family friends, who would bring him books on mathematics and scientific subjects. Particularly influential was his uncle, Jakob, an engineer with a good background in mathematics, who taught him the joys of self-study and discovery. Albert often

The Luitpold-Gymnasium (secondary school) in Munich, which Albert attended from 1888 to 1895. (Stadtarchiv München)

neglected his friends and playtime and found his greatest happiness and fulfillment in solving advanced mathematical problems.

But in 1894, when Albert was fifteen, his family, which owned an electrical engineering firm, left Munich for business reasons and moved to Milan, Italy. His parents decided that he should stay in Munich so that his education would not be interrupted, and he went to live with another family in the city. He disliked his school and its military-like discipline even more now that his family was gone and he was lonely for them. After half a year of anguish, he decided he could no longer bear it. Without first consulting his parents, he left school and took a train to Milan to be with them. He promised them that he would study on his own to prepare himself for the entrance examination at the Federal Institute of Technology (FIT; also known as the Polytechnical Institute, or Poly, and later as the ETH) in Zurich. This institution was one of the best universities in Europe for those interested in science and technology, and this is where Albert wanted to study. His parents took him at his word, allowing him to take the exam when he was only sixteen and a half years old, two years earlier than most candidates for admission. He passed it with high marks in mathematics and science, but the school administrators felt that he should gain a better knowledge of history and languages and become more mature before entering the school. He agreed to attend a high school in Aarau, Switzerland, a village near Zurich in the canton of Aargau for one year, 1895–1896, to become competent in those subjects and to be closer to the normal age of admission.

The Kantonsschule in Aarau, Switzerland, which Einstein attended during the 1895-1896 school year just before entering the Federal Institute of Technology in Zurich. (M. Schatzmann; Einstein Archives, Hebrew University of Jerusalem, Israel)

This last year of school before his university education began was wonderful for him in every way. The Swiss school system was much different from the authoritarian German regime he had detested, and the casual and friendly atmosphere suited his personality much better. The students were treated as individuals rather than as a military-like unit, and they were encouraged to be independent in both their thoughts and their actions. The teachers were not feared but respected, and students did not hesitate to go to them with their questions and problems. Switzerland was also a peaceable country that had no desire to go to war, conquer peoples, and gain new territory, even though its soldiers and army—which existed only for defense—were supposed to be among the best in the world. Switzerland was thus the ideal environment for a student of Albert's talents, sentiments, and character, and he thrived in his new surroundings. All of this was made possible by the Jost Winteler family, with whom he stayed during this school year and for whom he came to care deeply. Winteler, a warm and friendly man, was the popular headmaster of the school and the patriarch of a large family. One of the daughters, Marie, became Albert's first girlfriend, and one of the sons would become his sister Maja's husband a few years later.

The following year, 1896, Albert Einstein was

Maja Einstein, Albert's sister, ca. 1897, around the age of sixteen. (Courtesy of L. Monolith)

prepared to end his school days and boyhood and begin his college career. But first, at the age of seventeen, he decided to give up his German citizenship due to his dislike of the German military mentality. He also feared that he might be drafted into the German army. He remained without any kind of citizenship throughout his college years. In the fall, he enrolled at the Poly in Zurich and began his lifelong love affair with physics, mathematics, intellectual discussions, and higher education in general. The Poly, a highly respected technical institute of higher learning, had superior research facilities and a faculty that attracted students from many nations. This is still true today, though now it is called the Eidgenössische Technische Hochschule (ETH). Albert left Marie and the Winteler family behind him and rented a room in Zurich to start his life as a more independent young man in pursuit of his professional goals.

Easily bored and somewhat arrogant, the young Einstein did not particularly like to attend lectures on topics that did not interest him, and he would borrow his classmates' notes to study for exams. He formed close friendships with like-minded young men who would meet regularly in a small group they dubbed the "Olympia Academy." They would discuss physics and the problems of the day, and exchange lecture notes. He also took time out for romance with fellow physics student Mileva Marić, whom he admired for her intelligence and maturity. Mileva, a slight, dark-haired girl with a limp, came from Serbia and was four years older than he. They would study together and discuss their research interests, and Albert found in her the listener he desperately needed. She became the sounding board for his new ideas about physics. Before he finished his studies at the Poly, they had already decided to get married, and he announced their decision to his family the summer after graduation. His mother was bitterly opposed to the relationship, feeling that the shy, sullen Serbian girl with the sulky eyes was no match for her Albert. Still, Albert defended Mileva vehemently and promised to marry her once he found work after he finished his studies.

Upon graduation in 1900, Einstein faced problems finding a teaching position. Despite his brilliance, he had not been a favorite of the professors

Mileva Marić, Einstein's girlfriend
while at the Institute of Technology in Zurich, in 1896.
She later became his wife. (ETH, Zurich)

due to his lack of participation in the classroom. Therefore, he had a difficult time obtaining letters of recommendation from the faculty. This setback did not stop him from writing up his ideas and submitting his first scientific paper to a German physics journal, which published the paper of the unknown young scientist in 1901, when he was only twenty-two years old. During this time he also began work on a thesis that would lead to the Ph.D. degree at the University of Zurich. He received his Swiss citizenship and took on temporary teaching and tutoring jobs to earn subsistence pay, all the while remaining in touch with Mileva. Increasing his financial anxieties even further, Mileva became pregnant and gave birth to a daughter, Lieserl, in early 1902. That summer, after two years of being virtually unemployed, he was able to find work as a low-level inspector of new patents, mostly electrical inventions, in the federal patent office in Bern. He was hired there on a trial basis—if he proved to be a good worker, he could stay on and get a promotion. A few months later, his father died in Milan.

In January 1903, a year after Lieserl's birth and six months after he began his new job, Albert and

A street scene in Bern, Switzerland, around the time that Einstein found his first 'real' job in the patent office. (Author's collection)

Mileva were married in Bern and took up residence there. The record isn't clear, but it seems that Lieserl never lived with her parents, and Albert never even saw her. Mileva had returned home to Serbia to give birth and apparently left her there in someone's care. She went home again when she learned that the twenty-month-old Lieserl was sick with scarlet fever. After this time, no more mention is made of the little girl in any of the correspondence that has been left behind, nor has any further record of her, such as a birth or death certificate, been found. She was either given up for adoption or, more likely, died of her illness while living with Mileva's family, and for unknown reasons, was never spoken of again. This is a sad note in young Einstein's life.

In May 1904 Mileva gave birth to their son, Hans Albert. A few months later, Einstein's position at the patent office became permanent and he was given a promotion. The couple was able to settle down into a more relaxed and conventional life with their new baby boy. During this time and despite many financial and personal hardships, the young physicist came up with some of his most original ideas in physics. They were published in 1905 in five outstanding papers. Not only did he succeed in having his doctoral thesis published before he formally received his Ph.D.—during what came to be called his "year of miracles"— but he also published three of the most brilliant papers of his career, including the special theory of relativity. In addition, he published over twenty reviews of scientific papers published by others. Ten years later he came up with the general theory of relativity, published in 1916. Skeptical scientists did not accept all of his theories immediately, especially the German physicists who challenged them as "Jewish physics," but he defended them vigorously along with some of his supporters.

During the years 1906 and 1907, Einstein continued to work at the patent office, but he also continued to look for an academic position. He obtained a part-time job as a lecturer at the University of Bern in 1908 while he was also still examining patents. Finally, in 1909, he realized his dream and was appointed to a full-time position as professor of theoretical physics at the University of Zurich. After having made the patent office his professional home for seven years as a civil servant, the young man was about to embark on a lifetime in the academic world.

The young Einstein family moved to Zurich, where they found a comfortable apartment in a building that housed other friends whose children served as Hans Albert's little playmates. The word in the building was that the Einsteins kept a "Bohemian" household that was more unconventional than that of their middle-class neighbors in terms of furnishings and behavior. But even with his reputation as a rather awkward eccentric with a sharp tongue and shabby clothes, Einstein had risen in the esteem of his colleagues: that year, at the age of thirty, he received his first honorary doctorate—one of more than twenty to come—from the University of Geneva.

The year 1910 brought two joyous personal events. First, sister Maja was married in March to Paul Winteler. Maja had received a doctorate in romance languages from the University of Bern the year before, and she had remained in Bern, where her husband was a law student. A few months later, in July, the Einsteins' second child, Eduard, was born. Eduard was always a fragile child, and Mileva often took him into the Alps, believing that the mountain air was good for him, or that at least it would make his respiratory problems less serious. After overcoming the health problems of the first decade of his life, Eduard became an essentially healthy boy and young man who pursued a medical education. But at the age of twenty he developed schizophrenia, from which he suffered until his death. His dreams of becoming a doctor never came true. Unlike his father and brother, Eduard remained in Switzerland all of his adult life and therefore did not have much contact with his father, particularly after the Einsteins divorced and even more so after Einstein's immigration to the United States in 1933. Einstein was in touch with Eduard only through a friend, never writing to him directly. The year before Einstein died, he confided to his friend that he could not explain why he wanted no direct contact with Eduard, but he guessed that he feared his appearance in any form, even in a letter, would be painful for his son. In 1940 Einstein wrote to a friend that, since insulin injections were not successful in treating Eduard's illness, "I have no further hopes from the medical side. I think it is better on the whole to let Nature run its course." Eduard lived for another twenty-five years, dying in a psychiatric hospital in Switzerland in 1965 at the age of fifty-five, ten years after his father had passed away.

Einstein accepted an appointment as a professor at the German University of Prague in 1911. He resigned from his Zurich professorship and moved his family to the charming Czech capital. Here Einstein was hailed as a celebrity. The family settled into a spacious apartment that had enough room to house Mileva's mother as well as a maid and her daughter—providing a more conventional middle-class existence than they had been used to in Switzerland. But it wasn't long before Einstein was swamped with academic offers, among them the Poly in Zurich and the Universities of Vienna and Berlin. He accepted the offer from the Poly, and the Einsteins returned to Zurich in October 1912.

By this time, Einstein had already accepted his mother's view that his marriage to Mileva had been a mistake, and his marriage began to fall apart more quickly. He became reacquainted with his cousin

Ilse and Margot Löwenthal, Elsa's daughters and later Einstein's stepdaughters, in 1912.
(Courtesy of the Estate of Eva Kayser)

Elsa Löwenthal, who was a divorcée, three years older than he, and had two pretty daughters, Ilse and Margot.

For the next two years, Einstein and Elsa corresponded with each other while Einstein was living in Prague and Zurich and Elsa in Berlin. In spring 1912, shortly after seeing Elsa, the young man's fancy had already turned to love. "I must love *someone*. Otherwise it is a miserable existence," he wrote to Elsa. In the meantime, he wrote another famous paper on special relativity that year, which showed that the "*E*" in $E = mc^2$ was originally an "*L*."

Toward the end of 1913, Einstein was given a professional offer he couldn't refuse: a research position at the University of Berlin, without teaching obligations, and the directorate of a new institute of physics. This was a chance to pursue his

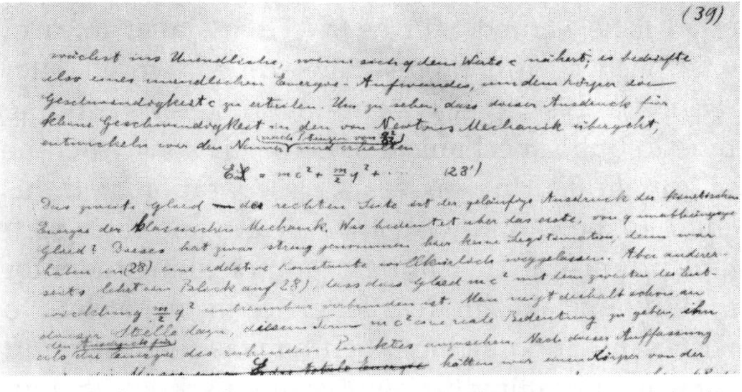

A page from a 1912 manuscript on special relativity, showing $E = mc^2$ in Einstein's own handwriting. The equation shows that E was originally an L.
(Albert Einstein Archives, Hebrew University of Jerusalem, Israel)

research without the interruptions of teaching duties, which he did not particularly like. Most important to him, it would give him the opportunity to mingle with the foremost physicists of his time. He resigned his position in Zurich and in spring 1914 moved his family, including Mileva, to Berlin. Realizing the marriage had essentially ended, Mileva returned to Zurich, taking the boys with her. In August 1914, World War I broke out.

Because of the war, border crossings became difficult, and Einstein was unable to visit his sons as often as he wanted. Still, he tried to maintain a warm relationship with them. Hans Albert would later recall that his father was quite a tinkerer, spending time to build mechanical toys for his sons when he saw them. During his years of separation from his children, he would write them warm and caring letters, promising to visit them, admonishing them for spelling mistakes in their letters or for neglecting to write, and announcing when he was sending a gift package. Recalling his own years of schooling, Einstein reminded Hans Albert that he shouldn't worry about his grades, that it's not necessary to get good marks in everything to succeed later in life. He

Einstein's sons, Eduard and Hans Albert, in 1918,
when they were eight and fourteen years old, respectively.
(Leo Baeck Institute)

would also remind him to practice his piano lessons and that he and Eduard should brush their teeth every day: "This is *very* important, as you will realize later on." He wrote them that he would try to spend one month of every year with them, so that "you will have a father who is close to you and can love you. You can also learn a lot of things from me that no one else can offer you so easily." He wrote Mileva that he needs this time with Hans Albert, who was now in his teens, to teach him to think objectively and to have an intellectual and esthetic influence on him. Yet, on the other hand, he also seemed content to leave the boys' upbringing in Mileva's hands so that he could concentrate on his professional activities—and on Elsa. Once the boys were out of their teens and Einstein became a celebrity, his visits to them became even less frequent, and the correspondence shows a growing resentment on the part of the sons.

The year 1914 was not a good year to be in Germany because the world had gone to war and Germany was in the middle of it. Germany, Austria-Hungary, and Turkey comprised the Central Powers, and much of the rest of the world, including

Japan, was allied against them. At this time, Einstein entered the pacifist and internationalist phase of his life, which put him at odds with many of his colleagues. He opposed the war and the fanatical military monarch, Kaiser Wilhelm II, who became a cultlike figure in Germany and was blindly obeyed by his equally fanatical army —a phenomenon that is not unknown in today's world, either. When

A soldier in Kaiser Wilhelm's German Army during World War I, ca. 1914. Einstein detested the war and the German military mentality.
(Braunsfurth family photo; author's collection)

ordinary citizens stood in the streets and cheered the soldiers as they marched in heavy goose-steps out of town, he couldn't believe his eyes. Such behavior was totally foreign to him and his beliefs.

Before long, ninety-three German scientists and intellectuals signed a proclamation trying to convince the rest of the world of the righteousness of the German cause, and Einstein became outraged. Along with a friend, he wrote an opposing pronouncement in 1915, upholding the values of a united European culture and stating that the war was insane and did great damage to human progress and science. Even in his later years he admitted that, although he tried to be universal in thought, he was still European by instinct and inclination. But this call for reason, his first political statement, fell on deaf ears, since Einstein did not yet have the celebrity and clout of his later years. Disappointed and frustrated, he devoted himself to finishing his work on general relativity and published it the following year, along with three papers on quantum theory.

In 1917 Einstein fell ill with a liver ailment and ulcer and was required to convalesce at home. The cheerful and motherly Elsa was only too happy to attend to him, a task made easier when he moved into her apartment. During this time, the British astronomer and physicist Sir Arthur Eddington received a copy of Einstein's paper on the general theory of relativity. Intrigued, he decided to get together a team of photographers to take photographs of the next solar eclipse, due in May 1919, to see if starlight is actually bent when it is passed by the Sun, as predicted by the general theory; one would be able to see such slight movement of the stars only when the Sun is darkened by the Moon during a total eclipse. While Einstein was recuperating from his illness, the Kaiser's power and strength had begun to wane, and by November 1918 he surrendered and abdicated his throne.

The following year, 1919, was an important one for Einstein in several ways. In February he was finally divorced from Mileva. Three and a half months later, on May 29, Eddington took his photographs of the solar eclipse. Four days later, Einstein and Elsa were married. Soon after that, the photos of the eclipse were developed, and news spread quickly throughout the world that they confirmed the general theory of relativity. Einstein became an instant global celebrity and forever lost the privacy he came to cherish in his later years. At first he was amused and enjoyed the novelty and adulation, but then he became bewildered and amazed by it. "Just as with the man in the fairy tale who turned whatever he touched into gold, with me everything is turned into newspaper clamor," he wrote to a friend in 1920.

In 1919 Einstein reinstated his German citizenship to show his respect for his German colleagues, but he retained his Swiss citizenship as well. Late in 1919 he had his first contacts with Zionists, whose goal was to establish a Jewish home state in Palestine. Einstein lent his name to their cause, though his view of Zionism was cultural rather than political. "Zionism, to me, is not just a colonizing movement to Palestine. The Jewish nation is a living fact in Palestine as well as the diaspora, and Jewish feelings must be kept alive wherever Jews live," he wrote in 1921. His Zionism, he said, did not exclude internationalism. His thoughts always remained global rather than nationalist, no matter where he lived or whom he supported.

The next year Einstein's mother died in Berlin.

The following year he won the Nobel Prize in physics; he gave his monetary award to Mileva to support his sons.

From this time on, Einstein's contributions to the world became more and more nonscientific. As a celebrity, he was able to influence the causes that were of interest to him, such as pacifism, world government, and Zionism. He traveled the world for the next thirteen years, giving both scientific lectures, and interviews and speeches on his thoughts and beliefs on various political and humanitarian matters. He traveled to the United States, Japan, China, Palestine, Spain, and South America, as well as England and other European countries. He kept up a lively correspondence with friends and colleagues all over the world, leaving a permanent record of his activities and interests. His home life was stable, with Elsa providing him with the kind of undemanding companionship he needed, mostly as a hostess at home and as a companion to social events and while traveling. She once again nursed him back to health when he fell ill again, this time with a heart problem, which weakened him for a year. During this time, both Ilse and Margot were married, as was

Hans Albert, and Helen Dukas was hired as Einstein's secretary and housekeeper. Dukas stayed in his household for the rest of his life, then became the archivist of his collected papers after he died.

Things changed in 1933: the Nazis came to power in Germany in January, and, as a Jew, Einstein's life was in danger. He resigned from his position in Berlin, denounced Germany, gave up his German citizenship, spent nine months in various parts of Europe while fleeing the Nazis, then embarked for the United States in October with Elsa and Helen Dukas. Ilse, Margot, and their spouses stayed behind.

In the United States, the Einsteins settled in the university village of Princeton, New Jersey. Here Einstein had been promised a position on the faculty of the newly established Institute for Advanced Study, which was housed at Princeton University until its own campus was completed in 1940. In 1934 Ilse died after a long illness while still in Europe, in Paris; then Margot and her husband, whom she later divorced, came to live in Princeton. Two years later, just before Christmas, Elsa died after a long battle with heart and kidney disease.

During these years that led up to World War II,

Einstein was able to live quietly as he was undergoing so many changes in his life. The university's parklike grounds and beautiful ivy-covered old buildings provided the intellectual and cultural atmosphere the Einstein family cherished, and they were happily absorbed into the community but were given their privacy. In 1939 sister Maja came to live in Princeton as well, and now Einstein led his bachelor life in the midst of a houseful of unmarried, childless women. All of them lived in the house at 112 Mercer Street until their deaths.

That year, too, while he was vacationing on Long Island, Einstein was informed that experiments on the rare metal uranium showed that it might be a suitable starting material to make bombs that could give off major explosions through nuclear chain reactions. Early in August, Einstein and his friend Leo Szilard drafted a letter to President Franklin D. Roosevelt, explaining the danger that lay ahead if a nation such as Germany were to make use of this knowledge. Einstein signed the letter and gave it to the friend of a colleague to deliver to the president. One month later, Germany invaded Poland and World War II started in

Europe. But it was not until October 11 that Roosevelt had found time to receive the messenger with the letter. The president had already met Einstein at the White House several years earlier and, with great respect, took his words seriously. He immediately instructed his officials to take action to further investigate, over the next year, the possibilities of the element uranium. A few months later, Einstein wrote a second letter stating the urgency of the situation now that Germany was totally engaged in warfare, reminding Roosevelt that German scientists might be producing an atom bomb for the war effort. Roosevelt was convinced that the United States must set up a program for the development of such a bomb. This project was implemented in December 1941, right after America entered the war—an effort referred to as the Manhattan Project and located at a highly secret laboratory in Los Alamos, New Mexico.

Ironically, Einstein could not take part in this project. Though he had become an American citizen in 1940, he was not able to gain the necessary security clearance from the FBI. Some members of the postwar U.S. government, who at the time were suspicious of those who dared to question their policies, did not like his "radical" activities on behalf of liberal-minded organizations whose stated goals were to work for world peace, understanding, and reconciliation among all nations. Because of these so-called un-American activities, they questioned whether Einstein could be a loyal American citizen. The FBI proceeded to gather a huge secret file on him, observing his every activity and keeping copious notes. Even Helen Dukas was suspected of being an informant. The agents finally admitted they were not able to pin anything un-American on the scientist. They didn't know that it was in fact Einstein who had tipped off the U.S. government about the possibility of building a bomb—before he was even a U.S. citizen. The highly questionable decision that Einstein was a security risk spared him any further involvement in the development of the bomb and, later, in its detonations. Therefore, he could honestly reflect later, "I have done no work on the atomic bomb, no work at all." Instead, he busied himself with humanitarian efforts to help get Jews out of Europe and to find positions for refugee scientists in America during and after the war.

Einstein officially retired from the Institute for Advanced Study in 1945 and became a pensioner, though he continued to have an office in Fuld Hall for the rest of his life. After the war, he supported the establishment of the state of Israel, spoke out for world government and disarmament, and continued to keep in touch with his many friends the world over. In 1948 Mileva died in Zurich; in 1951 Maja died in Princeton. In 1952, Einstein was offered the presidency of Israel, which he declined. On April 18, 1955, Einstein died in Princeton Hospital of an aneurysm of the abdominal aorta. Ten years later, his son Eduard died, followed by Hans Albert in 1973, Helen Dukas in 1982, and Margot in 1986.

Today, Einstein's progeny can be found in California, Switzerland, and Israel.

Einstein's Education

by Robert Schulmann

Einstein's personal interest in education was the subject of his very first public statement to a large audience on a nonscientific subject. In the middle of World War I, he wrote an emotional appeal to Germany to stop forcing children to take the difficult final exam required for high school graduation. Entitled "The Nightmare," the article appeared on Christmas Day 1917 in one of the largest newspapers in Germany. To understand what led to this outburst, it is useful to sketch out the course of Einstein's own education.

Albert began his public primary-school education in autumn 1885 in Munich, Germany. He was six and a half years old. A common practice at the time was to discipline schoolchildren with blows to the hand, and the young boy, who was just an average pupil, may have suffered his share of them. School authorities complained that little Albert stopped to think about the meaning of the teachers' questions rather than give quick responses. He would later compare these teachers to corporals in the army presiding over little soldiers. At home, things were better: he had to do his homework on time, but he accepted his parents' style of discipline more gracefully than that of his teachers. After completing his assignments, Albert would amuse himself with solitary games that required patience and precision, which would serve him well later in his research. His favorite pastime was constructing three-, even four-story houses of cards.

In autumn 1888, at age nine and a half, Albert entered the Luitpold-Gymnasium (secondary school) in Munich, which put him on the track toward a college education. Here the emphasis was on nonscientific subjects. In the forefront were the study of ancient Greek and Latin, subjects which did not interest him intellectually, but he performed well enough in both, especially Latin. One anecdote has it that his Greek instructor, angered by a poorly written assignment, predicted that nothing good would ever become of the boy.

Because he did not like the classes he had to take and the rote learning they required, Albert took matters into his own hands. When he was thirteen, before he was officially introduced to the subjects of algebra and geometry, he asked his parents to get him the mathematics text that would be used the following school year. His sister recalled that he spent countless hours arriving at proofs that differed substantially from those given in the book. In a matter of months, he succeeded in working his way through the entire mathematics program offered at the Luitpold school.

An older friend, a medical student who tutored

him, later recalled that already at their first meeting, the young boy "showed a particular inclination toward physics and took pleasure in conversing about physical phenomena." To supplement his unstimulating school readings on the basics of physics, Albert privately read a number of key textbooks, including a popular volume on the natural sciences, written in the middle of the nineteenth century. The topics in this book contain striking parallels to some of the budding genius's later ideas. For example, in the book, the author discusses the theme of atoms and the "secret" forces acting between them, which provided the background for one of Albert's earliest research programs: exploring the similarity between molecular forces and gravitation.

In summer 1894, when Albert was fifteen and his family moved from Munich to northern Italy, he remained behind to finish his studies at the Luitpold school. But he disliked it so much there that he managed to last only until the Christmas holidays. He was unable, he said, to stomach "the spiritless and mechanical method of instruction, which with my poor memory for words caused great difficulties, which appeared to me meaningless to overcome. For that reason I was prepared to accept any sort of punishment rather than learn to babble something strictly from memory." One teacher even complained that Albert's mere presence destroyed an atmosphere of respect in the classroom, which made him even more determined to leave and join his family in Milan at the end of 1894.

Albert told his parents that he would not try to finish his schooling in Italy, but he reassured them that he planned to prepare for a university career through self-study. He had set his sights on what was probably the most prestigious technical university in Europe, the Federal Institute of Technology (FIT) in Zurich, Switzerland. One can't help but admire this young man who was able to brush off the hostility of his German teachers and the uncertainty of his own parents and still aim so high. What makes his resolve even more admirable is the fact that, at age sixteen and a half, he was more than a year younger than the other students who took the entrance examination for the FIT. During this period of self-study, however, Albert acquired not only theoretical knowledge from advanced physics texts, but he also became interested in the practical concerns in the running of his father and uncle's Italian electrical factory. On one occasion he even helped to solve a difficult machine design problem that had stumped his uncle Jakob and other senior engineers.

When Albert took the entrance examination in Zurich in October 1895, he did well enough on the mathematical-physical part of the examination, but he failed the general portion of the examination, which tested his abilities in literary and political history as well as his proficiency in linguistic subjects. The director of the technical university advised Albert to attend an excellent secondary school in nearby Aarau, after which he would be able to enter the FIT automatically.

Albert spent the following school year, 1895–1896, at a school in Aarau, fifty kilometers west of Zurich. He had access to an excellent physics laboratory there. Even before coming to Aarau, Albert had been thinking about how light waves scattered through what was then called the "ether," an element that was said to fill the upper regions of space. While there, he started to think about what would happen if one pursued a light ray at speeds approaching the velocity of light, his "first immature thought-experiment that dealt with the special theory of relativity." All in all, he found his time in Aarau to be a welcome contrast to his experiences in Munich: "This school with its liberal spirit and teachers with a simple earnestness that did not rely on any external authority, made an unforgettable impression on me. In comparing it with six years schooling at an authoritarian German Gymnasium, I was made acutely aware how far superior an education that stresses independent action and personal responsibility is to one that relies on drill, external authority and ambition."

The reader may also be interested to learn that the myth of Albert's poor grades in secondary school may have had its origin in Aarau. In his first two terms there, he received grades of "1" and "2" in physics, algebra, geometry, chemistry, history, and music; in his last term, however, he received grades of "5" and "6" in these same subjects. Some have interpreted this to mean that Albert was, at best, an inconsistent pupil, or, at worst, a failure. Nothing of the sort: the explanation is quite simply that grade ranking at the cantonal school was reversed at the beginning of the school year, 1896–1897, so that "6" now became the best grade, which had been "1" only shortly before. Albert's final grades at the Aarau school were outstanding, and he easily passed his graduation exam.

Albert entered the FIT in October 1896 and graduated in July 1900 with a certificate that allowed him to teach mathematics and physics at the secondary-school level. Two themes characterize his university education, much as they had marked his earlier schooling: he refused to follow a prescribed curriculum, and he attached importance to the connection between the theoretical and the practical—an understanding he could gain without having to memorize endless lists and facts.

Describing himself later as a mediocre university student, Albert was at pains to define the attributes of a good student: "Facility in comprehension . . . , a willingness to concentrate on all that is presented [in the classroom], a love of order in order to take written notes on that which is offered in the lectures and then to elaborate on them in a conscientious manner. As I was forced to acknowledge, all these personal traits were thoroughly absent in my case." Instead, he learned to live with his guilty conscience and to organize his studies in such a way as to accommodate "my intellectual stomach and my interests." He followed some lectures with great interest, but for the most part he avoided going to class and studied the masters of theoretical physics at home. He also was clever enough to study for final exams by borrowing the notes of a friend who had attended all the lectures.

Though not particularly gifted as an experimentalist, Albert did embrace the practical side of physics during his university years and participated in a number of experiments. His emphasis on the unity of the theoretical and the practical is especially evident in his doctoral thesis, which combined a single theoretical claim—the existence of molecules—with a description of the law governing their behavior, shown by experimental data.

In 1900, at age twenty-one, Albert completed his studies, but now he had a difficult time finding a job. For two years he desperately looked for full-time work, but was only able to obtain part-time positions as a schoolteacher in the villages of

Schaffhausen and Winterthur. Finally, he was fortunate enough to find a job as patent clerk in the Swiss capital of Bern. This position, which he held from 1902 until 1909, allowed him to balance his work in the practical evaluation of patent models (mostly of electrical devices) with enough free time to think about his theoretical interests.

When he began to teach as an assistant professor at the University of Zurich in 1909, Albert began trying to apply in his classrooms the lessons he had drawn from his own education. But he found that he did not particularly enjoy teaching nor was he very good at it. He found that the kinds of subjects he taught needed the kind of structure he had disliked in his own education.

We have now come full circle to Einstein's first published statement on a nonscientific subject, his condemnation in 1917 of the compulsory secondary-school exit examination in Germany. After the description of Albert's own education, we may be in a better position to understand why he wrote the article. He attacks the examination on two grounds: (1) that it is useless for the pupil, and (2) that it is even harmful. It is useless because a teacher can judge a pupil's maturity and ability much better over the extended period of time that he or she has instructed the student than on the basis of a final examination for which the student prepared only quickly. It is harmful for two reasons: (a) the student is afraid of taking the examination because his future depends on it, and (b) the effort involved in committing to memory a large body of material can considerably damage the health of the pupil and often leads to nightmares in later life. The second aspect lies in the fact that the level of instruction sinks markedly in the last years of a pupil's schooling. As teachers concentrate less on the subject matter and more on impressing others with the brilliance of their pupils, they come to rely more and more on external drill than on acquiring knowledge for its own sake.

Einstein concludes his article by pleading: "For these reasons, away with the final examination!" To his mind, such drill-and-grill procedures destroy a pupil's curiosity and sense of individuality, the most precious gifts that an education can and should nurture and reinforce.

An Einstein Picture Gallery

Earliest known photo of Albert Einstein, ca. 1882 at age three. (Albert Einstein Archives, Hebrew University of Jerusalem, Israel)

Albert with sister, Maja, in 1884.
(Lotte Jacobi Archives, University of New Hampshire)

Class picture taken in 1889, when Albert was ten years old. He is in the bottom row, third from the right. (Stadtarchiv Ulm)

Albert Einstein at age fourteen in 1893.
(Albert Einstein Archives,
Hebrew University of Jerusalem, Israel)

Einstein with classmates at the Aargau Kantonsschule in his senior year, 1896. He is seated on the bottom left.
(ETH Bibliothek, Zurich)

Einstein as a college student, ca. 1900. (Lotte Jacobi Archives, University of New Hampshire)

Einstein with a baby, perhaps his son Hans Albert, ca. 1905. (Courtesy of L. Mondlith)

Einstein as a patent clerk in Bern in 1905. (Lotte Jacobi Archives, University of New Hampshire)

Einstein's image, among others, in the archway of the door to the Riverside Church in New York City, completed in 1930. He was the only living person, Jew, and scientist so honored. (Photo by the author)

With Charlie Chaplin at the premiere of City Lights, Los Angeles, 1921. (Leo Baeck Institute)

Albert and Elsa Einstein at the Grand Canyon, February 28, 1931, with a group of Hopi Indians, who nicknamed him 'the Great Relative.'

(Albert Einstein Archives, Hebrew University of Jerusalem, Israel)

Holding an Einstein puppet, during a visit to the California Institute of Technology in Pasadena, 1932.
(Archives, California Institute of Technology)

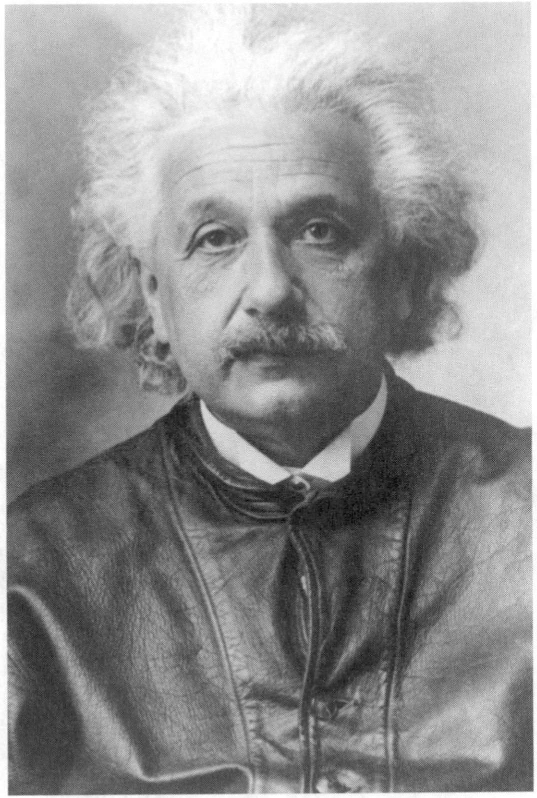

Einstein in 1935 in Princeton.

(Courtesy of Department of Physics, Princeton University)

At a summer vacation on Huntington, Long Island, in 1937. Einstein was an enthusiastic pipe smoker. (Lotte Jacobi Archives, University of New Hampshire)

Sailing at Huntington, Long Island, 1937. (Lotte Jacobi Archives, University of New Hampshire)

On the beach with daughter-in-law Frieda and grandsons Bernhard and Klaus, ca. 1937.
(Courtesy of L. Monolith)

In Princeton in 1938.
(Lotte Jacobi Archives, University of New Hampshire)

Einstein in 1938 in his parlor in Princeton, ready to leave for an appointment in New York.
(Lotte Jacobi Archives, University of New Hampshire)

In his study in Princeton, 1938. (Lotte Jacobi Archives, University of New Hampshire)

WORLD OF TOMORROW HEADS INTO THE BLACK (NEWS foto) Densely packed thousands jam World's Fair Court of Peace yesterday for ceremonies dedicating Palestine Pavilion. Huge Sunday crowd sent attendance totals soaring. By 11 P. M. the turnstiles counted 290,740 persons. This swelled total since Fair opening past 5,000,000 mark. World of Tomorrow officials indicated financial success of exposition seems assured.

Einstein with the women of his household and an unknown youth and elderly woman at the New York World's Fair, 1939. Left to right: Stepdaughter Margot, secretary Helen Dukas, Einstein, and sister Maja Einstein, who is said to have looked just like him. (Albert Einstein Archives, Hebrew University of Jerusalem, Israel)

With girls in a gym or dance class, late 1930s.
(Courtesy of L. Monolith)

Einstein and Margot receiving their citizenship in 1940 in Trenton, New Jersey.
(Courtesy of Todd Yoder)

Einstein relaxing in the early 1940s.
(Courtesy of Todd Yoder)

Enjoying a moment with baby boy John Steiding Jr., in 1946.
(Albert Einstein Archives, Hebrew University of Jerusalem, Israel)

In academic gown, on the day he received an honorary doctorate from Lincoln University, an African American institution, in Pennsylvania in May 1946. (Leo Baeck Institute)

Einstein with fuzzy slippers.
(Courtesy of Gillett Griffin, Princeton)

With baby J. M. Pietsch, 1951.
(Albert Einstein Archives, Hebrew University of Jerusalem, Israel)

With three little girls, ca. 1953.
(Albert Einstein Archives,
Hebrew University of Jerusalem, Israel)

The Letters

To Einstein's Wife and Sons, Hans Albert and Eduard ('Tete')

March 23, 1914

Dear Wife and Boys,

I hope you're well. I traveled well with Fokker [a colleague] and spent a day with my uncle Caesar, who sends you good wishes, in Antwerp. Tonight

I'll be at Ehrenfest's, where I'll be spending several days.

Tete should eat well!

Heartfelt greetings from your
Father

This postcard was sent from one of Einstein's many business trips, this one en route from Belgium to Holland. Eduard was sickly at this time and Einstein worried about his health, so he tried to cheer him up with the drawing of the little duck. (Original written in German.)

To Elisabeth Ley, Stuttgart

September 30, 1920

Dear Miss Ley,

Elsa tells me that you are unhappy because you didn't get to see your Uncle Einstein. Therefore I will tell you what I look like: pale face, long hair, and a modest paunch. In addition, an awkward gait, a cigar—if I happen to have one—in the mouth, and a pen in the pocket or hand. But this uncle doesn't have bowed legs or warts, and is therefore quite handsome; and neither does he have hair on his hands, as ugly men often do. So indeed it is a pity that you didn't get to see me.

With warm greetings from
your Uncle Einstein

This is the earliest extant letter from Einstein to a child, except for those to his own children. It is a good example of the tongue-in-cheek sense of humor he often displayed, also in his letters to adults. (Original written in German.)

From Children at a German Hostel for Working-Class Kids

March 1929

Dear Professor Einstein,

We send you our congratulations on your 50th birthday on March 14, 1929. And we hope that you will continue to live and work for a long time! We know that you are on the board of the children's hostels of the "Rote-Hilfe." We also know that it is because of the great love and support by friends of the children that we are able to spend ten glorious and happy weeks in the hostel.

With heartfelt greetings from all the children of the hostel "Mopr."

These children's hostels were supported by a humanitarian organization, the "Rote-Hilfe" ("Red Help," not to be confused with the Red Cross or Rotkreuz), which aided the working class in times of need. The hostels, some of them year-round "camps," gave the children of workers a chance to spend time in the countryside. As a birthday gift for the professor, these kids enclosed some of their drawings. (Original written in German.)

Herr Professor Einstein!

Wir gratulieren Ihnen zu Ihrem 50. Geburtstage am 14. März 1929.
Und wünschen, dass Sie noch recht lange leben und schaffen mögen!
Wir wissen, dass Sie im Kuratorium für die Kinderheime der Roten-Hilfe
Sind. Und wir wissen auch, dass es nur durch die grosse Liebe und
Unterstützung vieler Freunde der Arbeiterkinder möglich ist, dass wir zehn
So schöne, frohe Wochen im Kinderheim verleben können!

Es grüssen Sie herzlich alle Kinder aus
Arbeiterkinderheim „Mopr."

Postcards from Jewish Schoolchildren in Berlin for Einstein's Fiftieth Birthday

March 19 [1929]

Dear Mr. Professor,

I wish the professor a happy birthday. I hope you will be so kind as to come and visit the Jewish School on Greater Hamburg Street.

Many wishes,
Adolf

Original written in German in the old Gothic style. The drawings are a sample of many that were sent with the card.

19 März

Lieber Herr Professor.

Ich wünsche dem Herrn Professor zum Geburtstag viel Glück. Sie möchten bitte so freundlich sein und die jüdische Schule in der Großen Hamburger Straße besuchen kommen.

Viele Grüße

Adolf Simmel

From Arthur, Los Angeles

December 5, 1928

My dear Professor:

I have written an article on your celebrated Theory of Relativity, which I am enclosing, and would be more than happy to have your opinion on the paper enclosed herewith.

I would also like to know whether movement actually exists with its accompanied phenomena such as inertia, etc., in a space that contains no other object than the person or mechanism that is trying to get under motion? Would he not be stationary for the reason that if he did move, he would be no nearer nor further from anything?

Would you please favor me with the answer to this question as pertaining to the Relativity of the object in question with some other object in space. Relativity is a theory that has always fascinated me, and I have thought of it over and over again along with the great man who conceived it. Would greatly appreciate your opinion on my article.

Wishing you Godspeed, health, and happiness, I am

Sincerely yours,
Arthur, Age 12 years

To Arthur, Los Angeles

December 26, 1928

Dear young man,

In your article you were quite right in stating that motion can be experienced and presented by us only as relative motion. The Ancients already knew this, and it was also acknowledged by the opponents of relativity theory. But until the presentation of the general relativity theory, it was thought that the concept of absolute motion was necessary for the formulation of the laws of motion. To disprove this became an obstacle in relativity theory.

Your question asking what the world would be like if only one body were in it cannot be answered with certainty today, because we don't know if there could be space around that body. But we do know for sure that it is meaningless to speak of its motion.

But for you it would be better if you began to teach others only after you have learned something useful yourself.

With friendly greetings,
Albert Einstein

The last sentence was not meant as a harsh rebuke—only a slight one! Einstein was in general quite fond of curious children. This letter was written in German, but perhaps someone translated it for Arthur before it was sent. The New York Times *of February 10, 1929, ran a story on this letter.*

To a Young Man

January 13, 1930

Dear little cousin,

You are not the most savvy little customer, but it's good that you're at least a curious young fellow. So then:

The soup doesn't cool down as much because the layer of fat on the top makes evaporation more difficult and thereby also slows the cooling.

The matter of the lantern would be very serious if the waves in the line of sight were privileged. The cause of the phenomenon is due more to the fact that in the case of waves that come in under a shallow angle, the angle of rotation of the reflected wave is much larger when one rotates a mirror around an axis perpendicular to the line of sight than for rotations around an axis in the line of sight.

The phenomenon of the third riddle is due to the circulatory movement of the water, which is contingent on the fact that the centrifugal force due to friction on the bottom is less important than in the higher layers.

Cordial greetings from your
Einstein

The original letter was written in German to an unknown child.

To the Schoolchildren of Japan

[fall 1930]

In sending this greeting to you Japanese schoolchildren, I feel I have a special obligation to do so. For I myself have visited your beautiful country, its cities, houses, its mountains and forests, from which Japanese youngsters derive a love for their homeland. On my table lies a large book full of colorful pictures drawn by Japanese children.

If you are indeed receiving my greetings from so far away, remember that ours is the first era in which it has been possible for people of different nations to conduct their affairs in a friendly and understanding manner. In the old days, peoples spent their lives fearing and even hating one another because of ignorance on all sides. May the spirit of brotherly understanding among nations continue to grow. With this sentiment, I, an old man, send you Japanese schoolchildren my greetings from afar, with the hope that your generation will some day put mine to shame.

Albert Einstein

From Kenichi, Tokyo

October 7, 1930

Dear Sir:

Upon reading your most gratifying letter, we were completely beside ourselves with joy. Just imagine what a great influence your message will have on the Japanese children when it appears in our student newspaper, and how utterly thrilled they will be. Your valued opinion will stir their innocent hearts not only now, but it will continue to be a reminder for future generations.

We want to inform you that the editors of Shojon Club, the sister organ of Shonen Club, have asked your wife for a message and are eagerly looking forward to a reply.

We would like to present you with a gift of the book *Emperor Meiji,* which was published by us, and we will always make every effort to answer any questions you may have about Japan.

With profound respect,
Kenichi
Editor of Shonen Club

Einstein had visited Japan and other countries of the Far East in 1922. During this trip he learned that he had won the Nobel Prize in physics for 1921. He kept a travel diary of the trip in which he expressed his admiration for Japanese culture.

From Edward, New York

November 12, 1930

To Professor Einstein,

I am a student at the textile institute in New York. I hear that you are coming to America next month. We would like to invite you to a Seder at our school. It would be a great honor from a great man to have him at our school. I hope that you will accept this invitation. Our school is now at 210 E. 42nd St. After February it is moving to 18th St. and 9th Ave., New York.

With greetings,
Edward

Edward wrote his letter in German. Einstein visited New York in December 1930 before going to California for three months, but there is no record of a visit to the textile institute. For those not familiar with the word "Seder," it is a Jewish home or community service that includes a ceremonial dinner and is held on the first evening of the Passover holidays to commemorate the Jews' flight from Egypt.

From Esther, Freehold, New Jersey

April 15, 1935

Dear Professor Einstein:

In our Junior Business Training Class we are studying vocations. I am about to decide mine, but before I do so I would like to know how other people became famous.

Do you enjoy doing your work?

How did you come to choose it?

How did you become famous?

I suppose you meet many other great people too. Have you an ambition or enthusiasm for something else?

Please answer.

Respectfully yours,
Esther

Einstein came to live in the United States in October 1933. This is the first extant letter received from a child when he lived in Princeton, New Jersey.

From Phyllis, New York

The Riverside Church

January 19, 1936

My dear Dr. Einstein,

We have brought up the question: Do scientists pray? in our Sunday school class. It began by asking whether we could believe in both science and religion. We are writing to scientists and other important men, to try and have our own question answered.

We will feel greatly honored if you will answer our question: Do scientists pray, and what do they pray for?

We are in the sixth grade, Miss Ellis's class.

Respectfully yours,
Phyllis

Maybe Phyllis was encouraged by her Sunday School teacher at Riverside Church to write to Einstein because Einstein's image, among that of others, was sculpted into the arch over the entrance of the church. He was the only living person to have had this honor, and he was said to have been quite pleased with it.

To Phyllis, New York

January 24, 1936

Dear Phyllis,

I will attempt to reply to your question as simply as I can. Here is my answer:

Scientists believe that every occurrence, including the affairs of human beings, is due to the laws of nature. Therefore a scientist cannot be inclined to believe that the course of events can be influenced by prayer, that is, by a supernaturally manifested wish.

However, we must concede that our actual knowledge of these forces is imperfect, so that in the end the belief in the existence of a final, ultimate spirit rests on a kind of faith. Such belief remains widespread even with the current achievements in science.

But also, everyone who is seriously involved in the pursuit of science becomes convinced that some spirit is manifest in the laws of the universe, one that is vastly superior to that of man. In this way the pursuit of science leads to a religious feeling of a special sort, which is surely quite different from the religiosity of someone more naive.

With cordial greetings,
your A. Einstein

Einstein was often asked about his religious feelings and his belief in God, and his answers have been variously interpreted. But he said clearly that he did not believe in a personal God who can control the lives of others, one who punishes or rewards people. Rather, his was an attitude of cosmic awe and wonder, and a devout humility before the harmony of nature. Still, he was convinced that a superior intelligence reveals itself in the knowable world.

From Phyllis, Chambersburg, Pennsylvania

Camp Robin Hood [1937]

My dear Mr. Einstien,

I guess you have never received such odd letters before, but I hope you won't mind getting these. Yesterday we were talking about Mr. Houdini (who was the greatest magician in the world) and whether or not he knew the fourth dimension. Somebody said that you discovered the fourth demension and that it couldn't make you walk through walls. Pam Statton said that Mr. Houdini knew the fourth demension and that he walked through walls. Is this true? Please write and tell me if the fourth demension is time and explain it. Inclosed in this same envelope is a letter from Emily Jane Reese. She wants to know about it too. We felt awfully sorry for Mrs. Houdini and are trying to get in contact with her husband so as to help her get in touch with him, which she can't seem to do. Please don't think we're silly and all 'cause we aren't. I am

13 and not doing this for a trick or to make you seem foolish at all. Please believe us and write.

Yours truly,
Phyllis

P.S. Please excuse E. J. Reeses' spelling. Our counselor said that you ought to be called Professor so don't get mad if we address you as Mr. Einstein.

From Emilie Jane, Chambersburg, Pennsylvania

Camp Robin Hood

July 1, 1937

Dear Mr. Einstein:

I wish you could explain to me by letter how the fourth dimension works. One of the girls says that Mr. Hudini (a famous magician now dead about 10 years) could walk through walls just because he knew the fourth dimension. Immediately a "hot" discussion followed about Mr. Hudini. His wife (Mrs. Hudini) has tried very often to communicate with her dead husband. She has never succeeded so we thought we would help her by trying to communicate with him ourselves.

Outwardly we were all very brave about it, but inwardly our bones rattled. We placed a glass upside down in the middle of the floor. In the glass was a key. If the glass was gone or anything done to this glass or key, then we would know we were in contact with Hudini. We all went to sleep that night feeling very nervous. I was hoping that the glass would still be there in the morning. But in the morning, much to our horror the glass was *gone*!! One of the girls fainted, I got under the covers, and most of the girls hoped and prayed that Hudini wouldn't kill them. Just at this crucial moment a councellor told us that she had taken the glass for her use. Imagine our feelings!!

Signed—Emilie Jane, age 13

From Junior High School Einstein Club, New York

May 19, 1938

Dear Prof. Einstein,

Though none of us have been fortunate enough to meet you in person we have heard so much of you and your wonderful work as a scientist and mathematician that our mathematical teacher Mrs. Bernstein has formed a club for children interested in advanced mathematics, science or such things on this line and named it the "Einstein Club." Our club has just started this year and already we have about seventeen members, all very enthusiastic. We hope to continue this club next term and take up astronomy.

Every time we here of a new triumph of yours it is brought to meeting and discussed. Of course since we are still Junior High students we don't always readily grasp the ideas of your theories but as most of us are rapid advanced pupils we do get a pretty clear idea of many points with our teachers aid.

Dear Professor Einstein

Our club meets Wednesdays and today we found out your birthday was two days ago, so though we are a little late we do hope you had a Happy Birthday and will enjoy many more of the same.

Respectfully yours,
The members of the Junior
High School "Einstein Club"

To Schoolchildren

December 20, 1935

Dear Children,

With pleasure I can picture you children gathered all together during the holidays, united by a harmonious spirit instilled by the warm glow of Christmas lights. But also remember the lessons taught by the one whose birthday you are celebrating. This lesson is so simple, yet in almost 2000 years people still have not heeded it. Learn to be happy through the good fortunes and joys of your friends and not through senseless quarrels. If you allow these natural feelings to blossom within you, your every burden will seem lighter or more bearable to you, you will find your own way through patience, and you will spread joy everywhere.

With warm greetings,
your A. Einstein

This letter was originally written in German in response to a school's request for a meaningful Christmas message.

From John, Red Bank, New Jersey

April 10, 1937

My dear Dr. Einstein,

I am a student in the Red Bank Junior High School. Our class is particularly interested in our general science because we have been studying about astronomy. One day we were discussing "What holds the suns and planets in place?" and learned about Sir Isaac Newton's law of gravitation. Some students believe that your theory of relativity is an absolute contradiction of this law. We then decided to write you for your opinion. We are anxious to have this matter settled as soon as possible and I do hope that you will find time to give me a reply in the near future.

Very truly yours,
John

From Cyrus, South Africa

6th March, 1946(?)

Dear Albert,

I have just read a book which tries to explain your Theory of Relativity. I am sorry but I cannot understand it. But then I blame only myself as I am still a young school boy and have not advanced to your level of thought.

Please do excuse me taking this liberty of writing to you. But I always wanted to write to the Great Einstein. You shurly receive so many letters that this one may be discarded into the waste basket. But I do hope and beg that you will do me the honour of answering this letter. It is not the privalege of every school youth to possess a letter from Einstein. I am asking for a letter even in my prayers.

My reason for writing to you is to ask your opinion on a thing that has been worrying me of late.

Because of the Atom Bomb the world realizes that we must preserve peace at all costs. The best, and only way to do so, to my mind, is to accept universal tolerance based on equality of Mankind. And it is because of this that I am puzzled. I am growing up in a country which does not believe in the equality of all. The Natives in the country have no equal citizen rights and never shall. We all are brought up to despise the Natives here. I feel that this is a wrong attitude. But what solution can we find. Our history all revolves around the hating of the Natives. But this is only a small case in hand. What about the rest of the world. The white man thinks he is superior to the Asiatic, and the Asiatic in turn thinks he is better than the European. The Native thinks he is being suppressed and is waiting to overthrow the yolk of the white man.

Perhaps I am pessimistic. What do you think?

Please do answer.

Your admirer,
Cyrus

Though we have no response available, Einstein's writings indicate that he would have agreed with him. Among other instances, in 1946, when he received an honorary doctorate from Lincoln University, a university for African Americans in Pennsylvania, he gave a speech in which he lamented the fact that racism existed in America, too.

From Barbara, Washington, D.C.

January 3, 1943

Dear Sir,

For quite a while I have admired you. I have started to write before many times, only to tear the letter into bits. For you are such a brillant person & from what I have read you have always been so. I am just an average twelve year old girl in the 7A at Eliot Junior High School.

Most of the girls in my room have heros whom they write fan mail to. You and my uncle, who is in the Coast Guard, are my heros.

I'm a little below average in mathematics. I have to work longer in it than most of my friends. I worry (perhaps too much), although in the end I imagine it will all work out for the best.

One evening while our small family group was listening to the Radio Readers Digest I heard the brief story of an eight year old girl and yourself. I then told mother of my desire to write you. She said

"Yes" and that perhaps you would answer. Oh, sir, I do hope you will write. My name and address is below.

Sincerely yours,
Barbara

To Barbara, Washington, D.C.

January 7, 1943

Dear Barbara,

I was very pleased with your kind letter. Until now I never dreamed to be something like a hero. But since you have given me the nomination I feel that I am one. It's like a man must feel who has been elected by the people as President of the United States.

Do not worry about your difficulties in mathematics; I can assure you that mine are still greater.

Sincerely yours,
Professor Albert Einstein

The last sentence of this letter is quoted frequently. In developing his theories, Einstein always needed help with the sophisticated mathematics required to formulate them, and his collaborators included the world's best mathematicians. Still, he was better at math than most of us.

From Peter, Chelsea, Massachusetts

March 13, 1947

Dear Sir,

I would appreciate it very much if you could tell me what Time is, what the soul is, and what the heavens are.

Thank you,
Peter

From Fred, Newark, New Jersey

April 12, 1947

Dear Professor Einstein,

I am Fred, 9 year's old. I was born in Ulm. Did you ever hear of this town and from Buchau where my father came from? Please send me your signature. I collect stamps and want to start collecting signatures from famous people like you.

Yours truly,
Fred

From Charlene, Cambridge, Nebraska

May 9, 1947

Dear Mr. Einstein,

The seventh grade here in Cambridge, Nebraska read a story about your life yesterday. I truly believe everything about you that we read. If I'm getting *too* personal by asking you this you need not answer this question. Mr. Einstein, how old are you? You need not answer that question if you like. But I would like to know.

I would appreciate it very much if you would answer my letter, and would you please give me your personal autograph. If you have an extra photograph of yourself would you please send it to me? And please autograph your letter if you type it.

Thanks,
Charlene

I am in the seventh grade and I am 13 years old.

We know that Einstein was sixty-eight years old at the time.

Dear Professor Einstein

From Kenneth, Asheboro, North Carolina

August 19, 1947

Dr. Einstein,

We would like to know, if nobody is around & a tree falls would there be a sound, & why.

Please answer as soon as possible.

Truly yours,
Kenneth

Asheboro, N.C.
Aug. 19, 1947

Dr. Albert Einstein
Princeton University

Dr. Einstein,

we would like to know, if nobody is around & a tree falls would there be a sound, & why. Please answer as soon as possible.

Truly yours,
Kenneth

To Hans Albert

November 4, 1915

My dear Albert,

Yesterday I received your sweet little letter and it made me very happy. I'm very glad that you enjoy playing the piano. This and carpentry are, I think, favorite pastimes at your age, even more so than school. On the piano, play mainly the things that you enjoy, even if your teacher doesn't assign them to you. You learn the most from things that you enjoy doing so much that you don't even notice the time is passing. Often I'm so engrossed in my work that I forget to eat lunch. Also play ring toss with Tete—it can make you more nimble.

Kisses to you and Tete
from your
Papa

Regards to Mama

This letter is an excerpt from a letter written in German when Einstein was separated from his family. He was living in Germany while they lived in Switzerland.

From Don, Wewoka, Oklahoma

September 30, 1947

Dear Sir:

Our science class of Wewoka High School wishes to obtain information on the subject of Matter as we are composing our own booklet on this subject.

We would appreciate it very much, if you would send us some material on this subject. If this [is] not possible, perhaps you could give us a list of publications suitable for our age level.

Thank you very much for your consideration.

Sincerely yours,
Chairman of Committee
Don

From Mary, Auburn, Nebraska

November 17, 1947

Dear Mr. Einstein,

Our Algebra class have been studying all about you and your important works that you have been doing.

We made booklets all about you and your early life, your Nobel Prize, your teaching, and some of your theories. We were to have a picture of you so a lot of the pupils bought a Coronet Magazine just for your picture. We found pictures of you in Time, Life, and also all of the encyclopedia.

Out of all of the men that we studied, we found that you top our list because of your understanding look, your success, and you are better known than all of the rest (at least in Auburn), all of the rest looked mean and also like they thought they were successful but you look like a common ordinary guy.

We would like to know if you would please sign

your name to our paper so we could show the classes other than ours what we have done.

I would like to ask you to write me a letter but I suppose that would be asking altogether too much so please if you will just sign your name, but if you find time you might drop me a line or two.

I remain yours very truly,
Mary

From Tyfanny, South Africa

10th July, 1946

Dear Sir,

I trust you will not consider it impertinence, but as you are the greatest scientist that ever lived, I would like your autograph. Please do not think that I collect famous people's autographs—I do not. But I would like yours; if you are too busy, it does not matter.

I probably would have written ages ago, only I

was not aware that you were still alive. I am not interested in history, and I thought that you had lived in the 18th c., or somewhere around that time. I must have been mixing you up with Sir Isaac Newton or someone. Anyway, I discovered during Maths one day that the mistress (who we can always sidetrack) was talking about the most brilliant scientists. She mentioned that you were in America, and when I asked whether you were buried there, and not in England, she said, Well, you were not dead yet. I was so excited when I heard that, that I all but got a Maths detention!

I am awfully interested in Science, so are quite a lot of people in my form at school. My best friends are the Wilson twins. Every night after Lights Out at school, Pat Wilson and I lean out of our cubicle windows, which are next to each other, and discuss Astronomy, which we both prefer to anything as far as work goes. Pat has a telescope and we study those stars that we can see. For the first part of the year we had the Pleiades, and the constellation of Orion, then Castor and Pollux, and what we thought to be Mars and Saturn. Now they have all moved over, and we usually have to creep past the prefect's room to other parts of the building to carry on our observations. We have been caught a few times now, though, so its rather difficult.

Pat knows much more about the theoretical side than I do. What worries me most is How can Space go on forever? I have read many books on the subject, but they all say they could not possibly explain, as no ordinary reader would understand. If you do not mind me saying so, I do not really see how it could be spiral. But then, of course you obviously know what you are saying, and I could not contradict!

I must apologise once more if I have taken up some of your valueable time. I am sorry that you have become an American citizen, I would much prefer you in England.

I trust you are well, and will continue to make many more great Scientific discoveries.

I remain,

Yours obediently,
Tyfanny

CAPE TOWN.
South Africa,
10th July, 1946.

Dear Sir,

I trust you will not consider it impertinence, but as you are the greatest scientist that ever lived, I would like your autograph. Please do not think that I collect famous people's autographs — I do not. But I would like yours; if you are too busy, it does not matter.

I probably would have written ages ago, only I was not aware that you were still alive. I am not interested in history, and I thought that you had lived in the C18, or somewhere round that time. I must have been mixing you up with Sir Isaac Newton or someone. Anyway, I discovered during Maths one day that the mistress (who we can always sidetrack) was talking about the most brilliant scientists. She mentioned that you were in America, and when I asked whether you were buried there, and not in England, she said, Well, you were not dead yet.

In the British school system, "form" is the approximate equivalent of our "grade," such as "the twelfth grade" in school.

To Tyfanny, South Africa

August 25, 1946

Dear Tyfanny,

Thank you for your letter of July 10th. I have to apologize to you that I am still among the living. There *will* be a remedy for this, however.

Be not worried about "curved space." You will understand at a later time that for it this status is the easiest it could possibly have. Used in the right sense the word "curved" has not exactly the same meaning as in everyday language.

I hope that yours and your friend's future astronomical investigations will not be discovered anymore by the eyes and ears of your school-government. This is the attitude taken by most good citizens toward their government and I think rightly so.

Yours sincerely,
Albert Einstein

From Tyfanny, South Africa

September 19, 1946

Dear Sir,

I cannot tell you how thrilled I was to receive your letter yesterday. I still find it hard to believe that the most famous scientist in the world actually answered my letter! Thank you very much. The news that I had your signature went around the school in no time, and gave everyone something to talk about.

At the moment we are having a Mathematics lesson. The Maths mistress does not like our form, so we are meant to be working on our own. (She has refused to teach us as we talk too much.) It seems like the middle of Summer today; just the kind of day one cannot stay inside on. Outside birds are singing, and all that sort of thing, and here we sit learning that tan *d* is equal to something divided by something else! I wish I understood Maths, for it is needed in astronomical calculations, I believe.

I forgot to tell you, in my last letter, that I was a girl. I mean I am a girl. I have always regretted this a great deal, but by now I have become more or less resigned to the fact. Anyway, I hate dresses and dances and all the kind of rot girls usually like. I much prefer horses and riding. Long ago, before I wanted to become a scientist, I wanted to be a jockey and ride horses in races. But that was ages ago, now. I hope you will not think any the less of me for being a girl!

You can see the Southern Cross fine from the window of the room I have at school this term. I wonder if you have ever seen it. It is an awfully fine constellation, and when I am feeling fed up at night, after a day at school, I look at it, and it cheers me up no end. I have been lucky enough to see both Southern Cross and North Star, but I prefer our Southern Cross.

I say, I did not mean to sound disappointed about my discovery that you were still alive. In fact, I was just the opposite, for it's much nicer for one's favorite scientists in history to be alive, than to know that he died something like a century ago. I still wonder about space going on for ever, but I was very much encouraged by your saying I would one day understand the theory of curved space. I had almost given up hope that I ever would. I have been told that one must be pretty well advanced both astronomically and mathematically, to agree with such statements. I am afraid that from a theoretical point of view, my astronomy is just about on a par with my Maths. At the moment, that is. I hope to improve both some day.

There is not any news at school except that we won [against another school] in hockey last Saturday. Thank you once more, for your letter and signature.

Yours sincerely,
Tyfanny

To Tyfanny, South Africa

[Sept.–Oct. 1946]

I do not mind that you are a girl, but the main thing is that you yourself do not mind. There is no reason for it.

This is a transcribed fragment of a letter Einstein may have written by hand. The next letter from Pat mentions Einstein's letters *(plural), so he must have written more than one. We can imagine how thrilled Tyfanny must have been to receive two letters from Einstein, from so far away.*

From Pat, South Africa

March 4, 1947

Dear Mr. Einstein,

I suppose you will think this letter is from Tyfanny when you read the envelope, but I am one of Tyfanny's best friends, and I thought I would very much like to write to you as I am collecting an album of autographs of famous people. So far, all the autographs I have got are those of sportsmen so I thought I'd like to add yours to the collection and change the heading. I am one of the twins whom Tyfanny mentioned in her letter and, like her, I am very keen on science, especially astronomy.

I was hearing about your thoughts on Relativity a short while ago and I'm very interested in your ideas. I wonder if you could perhaps tell me the outline of your theory? Just a vague and brief summary, if you have any time, although, now I come to think of it, I'm pretty sure you don't have much time. I know that here at school we complain of having very little free time, so what it must be like beyond school, I can't imagine!

Tyfanny let my sister and I read your letters, as she was so thrilled about them and felt she would like to share her joy with some-one. They were very interesting letters and gave us terrific pleasure, so much so that we read them over and over again!

We have all been very excited lately over the Royal visit to South Africa and as they docked at Cape Town we had the honour of seeing them first and having them first set foot on our South African soil at our own Cape Town. My sister and I went to the Royal Garden Party—we were let off school for it—and we enjoyed it immensely. The Queen and princesses are so beautiful and gracious and the King is so friendly and charming that all South Africans have caught the "Royal fever"!

Who do you consider are the greatest scientists

alive today? I would very much like to get their autographs, but I am not at all sure who are the most outstanding. I would be very grateful if you could give me a few hints.

I remain,

Yours sincerely,
Pat

Einstein must have fulfilled his quota for writing letters to girls in South Africa, since Pat did not seem to get a reply. It's not clear which living scientists Einstein may have admired most, but he once indicated that among physicists, he admired Michael Faraday, H. A. Lorentz, James Clerk Maxwell, and Isaac Newton.

From A. J. Jr., Mullens, West Virginia

December 2, 1947

Dear Sir:

I was told by my science master about your therom

called the theroy of reality. I would like to have a copy of it if possible and convenient for you to send it to me. I am a scholar in high school and that is our study and I am very interested in my science work.

Thanking you so much,

I remain,
A.J., Jr.

From Doris, Perth Amboy, New Jersey

December 12, 1947

Dear Prof. Einstein,

I am ten years old, and I am collecting letters of famous people.

Would you please write to me and tell me your plans for the future.

Many thanks.

Sincerely yours,
Doris

From Shirley, New York

[1947/1948]

Dear Professor Einstein,

My name is Shirley and I am 10 years old. I'm writing to you because my cousin and I had a bet. He said you had 175 in entiledgence. And I said you had 190. Which one of us is right. Please write me at my address.

Your admirer
Shirley

My inteligence is 145. I would be very happy if you answered me. So please grant me the favor.

From Students at Hickman High School, Columbia, Missouri

December 30, 1948

Dear Dr. Einstein,

The homeroom students at David Henry Hickman Senior High School, Columbia, Missouri, have named themselves "The Little Einsteins." "The Little Einsteins" are trying to follow the high standards and ideals set forth by you by honoring each week a fellow student in our school who has done outstanding works or has unusual talent. We believe that this will recognize the ability of students and incourage them to do greater things.

You have been made an honorary member of this class. We take great pleasure in presenting you this Certificate of Award.

Little Einsteins of 106

A number of schools had Einstein clubs for children interested in the sciences.

From Richard, Pearl River, New York

January 1, 1949

Dear Professor Einstein,

I was so sorry that you were sick. You are a nice man and I hope you are better pretty soon. I learned about you in a comic book about atoms and I heard you were sick on the radio.

Best wishes and good health to you and your family,

Richard, Age six years old

From Edith

[January1949]

Dear Professor Einstein,

If you can spare the time and energy we would love to have your autograph, or even a scrap of paper from your waste basket. We would frame your autograph along with the eight pictures that we have of you.

Sincerely,
Edith

To Schoolchildren

[1949]

Dear Children,

Your gay pictures and messages have given me much pleasure. Thank you and good wishes,

Yours,
A. Einstein

Late in 1948, Einstein was diagnosed with an aneurysm of the abdominal aorta and received many get-well messages from adults as well as children.

From Heidi, Halle, Germany

March 14, 1949

Dear Professor,

Today we celebrated my birthday. I became eleven years old. In the afternoon, Daddy read *Roland of Berlin,* which the mailwoman brings to us every week. This time it is written in there that today is your birthday as well, and you turned 70 years old. And because we celebrate our birthdays at the same time, I am sending you belated good wishes from all of us. I have two sisters. My daddy is allowing me to do this, but he will only help with the address. Next year I will have to write earlier so that the letter will arrive on time. We have an aunt who is a seamstress in Hoboken near New York. It takes 4 weeks to send a letter there. And my daddy says you are a famous man and that we'll soon study you in physics class, and I should be flattered that we share the same birthday.

Best wishes,
Your Heidi

The original letter was written in German.

From a Third-Grade Class in Michigan

March 14, 1949

Dear Dr. Einstein:

We hope you had a very happy birthday. We knew it was your birthday because we read it in a book about famous people. Our room hopes you got lots of nice birthday wishes.

Your friends in Third Grade.

This was one of many similar messages Einstein received from children all over the world for his seventieth birthday.

Tappan School
Ann Arbor Mich
March 14, 1949

Dear Dr. Einstein:

We hope you had a very happy birthday. We knew it was your birthday because we read it in a book about famous people. Our room hopes you got lots of nice birthday wishes.

Your friends in Third Grade.

MR. EINSTE
IN

From Sam, New Martinsville, West Virginia

May 6, 1949

Dear Mr. Einstein,

I am writing to you to settle an arguement another boy and I had in school today. We are both in the eighth grade. This is a rather unusual question but it concerns you very much. This friend of mine claims that every genius is bound to go insane because all geniuses in the past have gone insane. I could not make him believe that there ever had been a genius in the past that hasn't gone insane. I said that you were a genius and you hadn't gone insane. My friend said you would go crazy in a year or less. I said you wouldn't. Our teachers wouldn't take sides and it got to be quite a heated arguement and we decided to write to you and see what you thought about it. If at all possible try not to go insane at all. Confidentaly, I think my friend isn't quite all there.

Please write and give me your opinion on this question (whether or not you'll lose your famous and valuable mind).

Admiringly yours,
Sam

From Gloria, Alameda, California

October 4, 1949

My dear Dr. Einstein,

As secretary of the U.B. of P. (Alameda High School Physics organization), I have the great honor of advising you that you have been elected an honorary life member of this organization.

The members of the club take this means of expressing their great appreciation to you for the great work you have done in the field of physics and theoretical physics.

The members hope that you will accept this membership.

Respectfully yours,
Gloria
Secretary

Essay by a Child

[ca. 1949, Princeton]

A Memorable Day

One day two or three years ago my mother told my brother and me that the great scientist, Professor Einstein, was coming for lunch. When the day finally came my mother told us to be on our best behavior so that he would get a good impression of us. As the moment finally came, I began to get scared. I did not know how to act in front of a celebrity. But after he got talking to us I lost all my fear, because I knew he liked children.

I didn't think I would be able to eat any lunch that day on account of my excitement, and fear for that great man, but when I saw how much Professor Einstein ate and how he enjoyed his lunch, I also enjoyed mine.

After lunch he graciously asked if we would like him to play his violin which he had brought with him. Of course we were delighted. Then the great professor took off his coat and rolled up his sleeves, and played two or three pieces.

He never mentioned his work, but simply talked about things that we could understand. He was so simple and kind that one would never think that he was a great professor. In my mind he is not only great but noble.

Einstein Biography by an Unknown Child

[ca. 1949]

Einstein

Albert Einstein came from gremeny, he is a short *bald* man. He is one of the smartest men that we have in this country. Einstein's theory is too much for me. Well anyway he is a teacher at Princeton.

He is my cousin too, yep. He is 70 years old and he can do O.K. for himself. He lives at the collage.

He is a mathematician to. He was born in 1879 of jewich parents, he went to school both germany and Switzerland.

In 1933 Einstein left Germany because of Hitler. He is a quiet peacful little man. Who is much admired by everyone and me. He has a wife, but no children.

Surely this biography by a child was one of many written during Einstein's life, but only this one seems to have been kept in the archive. Maybe the claim that Einstein is the child's cousin is true, even though he or she didn't quite get all the facts right.

From Judith, Budapest, Hungary

February 1950

Dear Albert,

We—my friend and I—should like to correspond [with] someone in America. Well, because we have no girl's address we write to you a letter. Please, answer us, if you also would like [to] correspond, or give our address to someone, who is capable [to] correspond willingly.

We are 15–16 years old girls and live in Budapest the capital of Hungary. We don't know well English, for it is no our mother tongue and we learn it only since about two years. But we are hoping that you will understand our letters. We are going to the grammar school's first class (I think, in your country it is the 5th class). At school we learn mathematic, Hungarian literature, Russian language, chemia, history, geography, singing, drawing, and gynastics. My favorite subject is singing.

Do you like the music? I like very much. I play on piano since about 4 year and I like it so much. My hobbies are concerts, opera, theatre and interesting books. What is yours? I liked to go to cinema too, but nowadays there are no good films here.

We don't live inside the city, but outside among the mountains in a villa. So I must travell in tram about till one hour every morning to school.

I would like if you would write me answer. Please write me much about yourself and your family, and your home. Write me, that by what you are interested? Can you send a photograph? I also send you one of myself soon. Next time I shall write more.

I send my best regards to your family.

Hoping, I hear from you soon,
Judith

P.S. Please send a corresponding address to my friend Ildiko or me.

It's not clear, but this girl, in her broken but understandable English, does not seem to know she is writing to a famous scientist. Perhaps someone gave her "Albert's" name as a joke, and she thinks she is writing to a teenage boy.

To Monique, New York

June 19, 1951

Dear Monique:

There has been an earth since a little more than a billion years. As for the question of the end of it I advise: Wait and see!

Kind regards,
Albert Einstein

I enclose a few stamps for your collection.

Unfortunately, Monique's letter is lost, but we can imagine what she asked in it.

From Anna Louise, Falls Church, Virginia

February 8, 1950

Dear Prof. Einstein,

I have a problem I would like solved. I would like to know how color gets into a bird's feather. I have many beautiful Parakeets, which have many beautiful colors. I asked my father and he told me to ask you.

Sincerely yours,
Anna Louise

From Claire, Brooklyn, New York

February 15, 1950

Dear Mr. Einstein,

Please accept this humble letter of appreciation I am sending you. You are in my estimation the most wonderful man alive today.

My appreciation for what you have done for the American people is more than I can attempt to write on a few lines. But, I would like to say thanks. Thanks, for saving lives as you did and for helping my family as you did millions of others all over the world.

And thanks, thanks for using your intelligent mind to help me and millions of others like me. Thanks!

Sincerely yours,
Claire, Age 13-½

From a grateful Jewish refugee girl. Einstein helped many Jews to obtain visas enabling them to come to the United States in the 1930s and 1940s.

To Hans Albert

October 13, 1916

My dear Albert,

Your letter pleased me so much. I just received it this morning. [A friend] and I decided to go on a trip with you once peace has returned. I'm very glad that Mama is gradually feeling better. Prof. Zangger promises that she will get healthy again, but she needs complete rest for a while yet. I'm especially glad that both of you are already so understanding and independent that you can get by quite well with only the maid. But I'm sorry that you're not taking piano lessons anymore. How did this happen? It's at least as important to me as what you're learning in school. Don't worry about your marks. Just make sure that you keep up with your work and that you don't have to repeat a year. But it's not necessary to have good marks in everything.

I miss you and Tete very much and am very eager to see you both. Although I'm over here, you have a father who loves you more than anything else and who is constantly thinking of you and caring about you. Send Mama my kind regards, and fond kisses for both of you from your

Papa

From Frank, Bristol, Pennsylvania

March 25, 1950

Dear Dr. Einstein,

I want to know what is beyond the sky. My mother said you could tell me.

Yours truly,
Frank

Frank, Einstein would probably have told you that there is still more sky beyond the sky, in various layers of the atmosphere called the troposphere, the lowest layer in which most clouds occur, up to about ten miles; the stratosphere, which extends to thirty miles above sea level; the

mesosphere, up to fifty-five miles, where a marked decrease in temperature occurs; and finally the ionosphere, in which the temperature increases substantially.

From David

January 11, 1951

Dear Dr. Einstein,

We are three boys nine years old who are all interested in science. We decided to form a club dedicated to science. We have several other boys who have joined with us.

After taking a vote we decided to call ourselves "The Junior Einsteins" because we all admire you very much. And would like to follow in your foot steps.

May we have your permission to use your name? We would also like to have your picture for our club room.

Thank you very much.

Respectfully yours,
David and friends

From Michiyosy, Kagosima, Japan

February 21, 1951

Our dear physicist Mr. Albert Einstein,

A week ago, I found "The Evolution of Physics" and "Out of My Later Years" in our school's library. And I have just [read] this books with great eagerness. This books drew my interest. Then I began to develop a great liking for physics. "Here is a great problem to be solved," I said to myself, "I will be the man to solve it." Thank you very much for your choice that chose physics for me.

I approve all your thinking that all country in world ought united for human sacrifice. This thinking is approved by not only I but many boys and girls of Japan. Then they believe that it is right. Do not forget that there is many boys and girls that love peace and liberty.

Please introduce me your sons and daughter. I have just become 16-year-old.

Yours truly,
Michiyosy

From a Father, New York

ca. February 9, 1950

Dear Dr. Einstein,

Last summer my eleven-year-old son died of polio. He was an unusual child, a lad of great promise who verily thirsted after knowledge so that he could prepare himself for a useful life in the community. His death has shattered the very structure of my existence, my very life has become an almost meaningless void—for all my dreams and aspirations were somehow associated with his future and his strivings. I have tried during the past months to find comfort for my anguished spirit, a measure of solace to help me bear the agony of losing one dearer than life itself—an innocent, dutiful, and gifted child who was the victim of such a cruel fate. I have sought comfort in the belief that man has a spirit which attains immortality—that somehow, somewhere my son lives on in a higher world.

I have said to myself: "Is not everything in this universe created in accordance with a fixed purpose and does not everything accomplish its purpose? What would be the purpose of the spirit if with the body it should perish; of what benefit would be to us the faculties of thinking and reasoning if they should attain us no more than they do here on earth and the insight to the full truth be barred to us forever? . . . "

I have said to myself: "it is a law of science that matter can never be destroyed; things are changed but the essence does not cease to be. Take a quantity of matter, divide it and sub-divide it in ten thousand ways, still it exists as the same quantity of matter with unchanged qualities as to its essence and will exist even unto the end of time. Shall we say that matter lives and the spirit perishes; shall the lower outlast the higher?"

I have said to myself: "shall we believe that they who have gone out of life in childhood before the natural measure of their days was full have been forever hurled into the darkness of oblivion? Shall we believe that the millions who have died the death of martyrs for truth, enduring the pangs of persecution, have utterly perished? Without immortality the world is a moral chaos." . . .

I write you all this because I have just read your volume *The World as I See It*. On page 5 of that book you stated: "Any individual who should survive his physical death is beyond my comprehension . . . such notions are for the fears or absurd egoism of feeble souls." And I inquire in a spirit of desperation, is there in your view no comfort, no consolation for what has happened? Am I to believe that my beautiful darling child ... has been forever wedded into dust, that there was nothing within him which has defied the grave and transcended the power of death? Is there nothing to assuage the pain of an unquenchable longing, an intense craving, an unceasing love for my darling son?

May I have a word from you? I need help badly.

Sincerely yours,
R.M.

Einstein took time to answer the anguished father's letter with the following response.

To R. M., New York

February 12, 1950

Dear Mr. M.,

A human being is part of the whole world, called by us "Universe," a part limited in time and space. He experiences himself, his thoughts and feelings as something separate from the rest—a kind of optical delusion of his consciousness. The striving to free oneself from this delusion is the one issue of true religion. Not to nourish the delusion but to try to overcome it is the way to reach the attainable measure of peace of mind.

With my best wishes,
sincerely yours,
Albert Einstein

From Six Little Scientists, Morgan City, Louisiana

[1951]

Dear Proffesser,

We are six children who took an interest in Science. We are in sixth grade. In our class we are having an argument. The class took sides. We six are on one side and 21 on the other side. Our teacher is also on the other side so that makes 22. The argument is whether there would be living things on earth if the sun burnt out or if human beings would die. We are not going to say anything we don't believe. We are going to keep what we believe in until it is proved different. We believe there would be living things on the earth if the sun burnt out. Will you tell us what you think. We have some other questions we have been wondering about. If it wouldn't bother you too much could you please answer them? They are: Does the sun give off hydrogen? Are the stars bigger than the sun? Do we have a chance to become scientists? We are:

Linda, age 11	Richard, age 11
Brenda, " 11	Rosalie, " 11
Ubain, " 11	Glenn, " 11

We would like you to join our Six Little Scientists, only now it would be Six Little Scientists and One Big Scientist. Please give us each your autograph so we can use them for club badges and also to help us remember how you helped us and showed us something if we are wrong. Probably there are many things misspelled or done wrong on here because we are not showing this to the teacher. Our teacher's name is Mrs. Smythe.

If you would join our club you must not tell any of our secrets and you would be our special friend and not a proffesser.

Love and lollipops,
Six Little Scientists

P.S. Linda wrote the letter.

To Six Little Scientists

December 12, 1951

Dear Children:

The minority is sometimes right—but not in your case. Without sunlight there is:

no wheat, no bread,
no grass, no cattle, no meat, no milk,
and everything would be frozen.
No LIFE.

A. Einstein

From Ann

[1951]

Dear Mr. Einstein,

I am a little girl of six.

I saw your picture in the paper. I think you ought to have a haircut, so you can look better.

Cordially yours,
Ann

DEAR MR. EINSTEIN
I AM A LITTLE GIRL OF
SIX.
I SAW YOUR PICTURE
IN THE PAPER.
I THINK YOU OUGHT TO
HAVE YOUR HAIRCUT,
SO YOU CAN LOOK
BETTER.
CORDIALLY YOURS,
ANN

From Kiyokazu, Talatsuki-Shi, Japan

June 15, 1951

Dear our respectful Dr. Einstein:

Father of 20th century Scientists, Originator of the Theory of Relativity, Pioneer in Atomic Physics, Strong protector of peace and liberty in the world, whom we hold in high respect as our spiritual benefactor, with the best wishes—for your health and happiness—of all of us who study in a small school in a rural community of Japan.

Please crave for 'United Nations' that we desire for the peace of the world, and we desire that Koria in confusion will be governed well swiftly, and the peace conference will be held swiftly, and the treaty of peace will be concluded swiftly, too.

We shall be happy if you will accept some pictures and some picture postcards (by general mail on 30th May and 11th June) under separate covers.

Yours very truly,
Kiyokazu

From Ria, Holland

[ca. 1952]

Dear Professor Einstein,

From your secretary I received your address, that made me very glad and my hearty thanks for it! No, this letter is not an important letter for you, but *very* important! I have a fiery wish and I hope so, that you will not refuse it, for I should be you so thankful!!!

Do you want to write, please, your signature on the little card and then send to me, yes, Professor Einstein? Ah!! My hearty thanks! Alas I write in my letter very many mistakes and I am ashamed. But forgive me it, my age is fifteen years and I have still only three years English lesson at school.

Our teacher tells well once about you, and Professor Einstein I have got very much admiration for you. How eager I want to see you once and I hope it later, when I am a journaliste and possess much money (ha! I hope!).

This letter I send by airmail and you will receive it now fast. I desire to the day, that I shall come out of school and my mother will say me at home, "Ria, there is a letter for you, the signature from Professor Einstein!" Oh, I desire it!

Now I finish my letter for it is time to go to bed for me. I hope, that I have not asked you a little too much! In advance thanks you kindly a girl-friend named

Ria

From Jerry, Richmond, Virginia

[1952]

Dear Sir,

I am a high school student and have a problem.

My teacher and I were talking about Satan. Of course you know that when he fell from heven, he fell for nine days, and nine nights, at 32 feet a second and was increasing his speed every second.

I was told there was a foluma [formula] to it. I

know you don't have time for such little things, but if possible please send me the foluma.

Thank you.
Jerry

Maybe Einstein couldn't come up with the formula because of his own troubles with mathematics.

From John, Culver, Indiana

[1952]

Dear Dr. Einstein,

My Father and I are going to build a rocket and go to Mars or Venus. We hope you will go too. We want you to go because we need a good scientist and someone who can guide a rocket good.

Do you care if Mary goes too? She is two years old. She is a very nice girl.

Everybody has to pay for his food because we will go broke if we pay!

I hope that you have a nice trip if you go.

Love,
John

Dear Mr. Einstein
My Father and I are going to build
a rocket and go to Mars or Venus.
We hope you will go too! We want you
to go because we need a good scientist
and some one who can guide a rocket an
Do you care if Mary goes too? She is
two years old. She is a very nice gi
Everybody has to pay for his fo
because we will go broke if we pay!
I hope that you have a nice trip if
you go.

Love
John

From Curtis, Hardin County Science Club, Elizabethtown, Kentucky

May 20, 1952

My dear Doctor:

You once said in 1946, "Today lack of interest would be a great error . . ." Upon these immortal words we have founded a science club for future scientists of America. Our members number, at the present, 28; and range in age from 14 to 18 years. We have two very fine men as our sponsors: Dr. Rustin of the Hardin County Health Dept. and Mr Kerrick, local science teacher. Another guiding light has been Dr. William S. Webb, recently retired from head of the Physics Dept. at the University of Kentucky, and this year, president of the American Physics Teachers Association.

The members of this club, of which the constitution is enclosed, wish me to ask you a question which is very deep in their minds. Needless to say, they idolize you and reverence your name. We have often heard that you are *Dr. Einstein,* not to be bothered by trifles and unimportant matters. Yet we do not picture you wholly as such; rather we see you as America's most beloved and distinguished scientist, warm and cordial, the intriguing continental personality that you are. Now, recalling your "today lack of interest" statement, we wish to ask that you accept the honorary presidency of our organization. To these 28 young men and women, to these 28 leaders of tomorrow, your decision means either a life-long dream come true . . . or a shattered future.

We eagerly await your choice on this matter. And for your convenience I have enclosed a self-addressed envelope.

Respectfully yours,
Curtis, President

From June, British Columbia, Canada

June 3, 1952

Dear Mr. Einstein,

I am writing to you to find out if you really exist. You may think this very strange, but some pupils in our class thought that you were a comic strip character. Mr. Robinson our English teacher suggested we write to you because we were talking about you in class. One question I would like to ask is if you make any mistakes?

Are you interested in music? Would you please write back as soon as possible so I may receive the letter of yours by June 25, 1952, because that is when we get out of school. I am in grade 7, go to the Trail Junior High School.

Yours sincerely,
June

Einstein apparently didn't answer, but he did like music very much and it was a big part of his life. He played the violin and the piano, and enjoyed listening particularly to the music of Mozart, Bach, and Schubert. He often played in informal quartets and also as a member of groups playing for charity. By 1950 he gave up playing the violin and played the piano instead.

From Neil, San Diego, California

August 5, 1952

My dear Dr. Einstein,

My brothers and I were having a discussion on whether you consider yourself a genius; although the whole world knows you for one, I do not believe that you consider yourself as such.

Please answer.

Sincerely yours,
Neil

P.S. I am 14 and I am sure you will receive another letter from my younger brother who is 9. Would you please answer each of us. Thank you.

August 5, 1952.

My Dear Dr. Einstein,

My brothers and I were having a discussion on whether you consider yourself a genius; although the whole world knows you for one, I do not believe that you consider yourself as such.

Please answer.

Sincerely yours
Neil

From Peter, San Diego, California

August 5, 1952

Dear Dr. Einstein,

My brothers and I were having a discussion. One 16, the other 14, and my self 9 years old.

We were talking about you being a genius. The world and we know you are a genius, but do you think you are a genius?

We would all like to know if you consider yourself a genius.

Yours sincerely,
Peter

From the Brothers, San Diego

August 25, 1952

Dear Dr. Einstein,

My brothers and I thank you for your prompt reply on the question whether you consider yourself a genius.

We shall all treasure your answer.

Yours sincerely,
Neil and Peter

P.S. *We* are agreed that you are a "genius."

Unfortunately, there is no record of Einstein's reply. But it would be safe to guess that he wrote the boys that he did not consider himself to be a genius. Einstein was a modest man and thought he was only more curious about the world than others, and that curiosity is a childlike quality that most adults seem to lose.

From Friendship Among Children and Youth (Founded Under UNESCO), New York

November 18, 1952

Dear Prof. Einstein:

The First International World Friendship Festival and Seminar will take place in Salzburg from December 12th to 16th, 1952. About 150 teenagers from 14 countries will be the delegates.

The topic of the Seminar will be: "Children's and Young People's Active Participation in Building World Friendship as a Basis for Lasting Peace."

Our common dear friend, Dr. Elisabeth Rotten, will be there and help to create a concrete program for building friendship among the children and youth of the world.

A thought from you for the young people to take home with them would mean so very much. We would therefore be very grateful if you would be kind enough to send a message to this event. . . .

To Friendship Among Children and Youth, Salzburg, Austria

November 22, 1952

You young people should consider yourselves fortunate that you, in your impressionable years, have the opportunity to exchange viewpoints and ideas with those of a variety of cultural backgrounds. There is no better opportunity to acquire the lifelong insights that are necessary for the resolution of international problems and conflicts.

In the hope that your endeavors have a lasting impact, I send you my warmest greetings and wishes.

Albert Einstein

This letter, originally written in German, reflects Einstein's lifelong commitment to peaceful resolution of international conflicts and his belief that a world government was the only way to ensure world peace.

From Robert S., 5th Grade Teacher,
Westview School, Zanesville, Ohio

November 26, 1952

Dear Sir:

I am enclosing a few letters that my class has written to you exactly as they were given to me. It seems that most of these children were so shocked when they read that human beings have been classed in the animal kingdom. I have tried my best to explain why this is so, but even though the majority of children accept it as such, there are still a few who cannot reconcile the thought.

I can readily understand why such young minds are concerned about the idea. They have been wanting me to help them write letters to some great thinkers on the subject. Could you spare a few minutes of your most valuable time in answering the question for them.

Very respectfully yours,
Robert S.

One of the pupils' letters, representative of the others, is printed below. Since it's dated earlier, Robert must have been collecting the letters for a few days.

From Carol, Zanesville, Ohio

November 12, 1952

Dear Dr. Einstein,

I am a pupil in the sixth grade at Westview School. We have been talking about animals and plants in Science. There are a few children in our room that do not understand why people are classed as animals. I would appreciate it very much if you would please answer this and explain to me why people are classed as animals.

Thanking you,

Sincerely,
Carol

To the Children of Westview School

January 17, 1953

Dear Children:

We should not ask "What is an animal" but "what sort of thing do we call an animal?" Well, we call something an animal which has certain characteristics: it takes nourishment, it descends from parents similar to itself, it grows, it moves by itself, it dies if its time has run out. That's why we call the worms, the chicken, the dog, the monkey an animal. What about us humans? Think about it in the above mentioned way and then decide for yourselves whether it is a natural thing to regard ourselves as animals.

With kind regards,
Albert Einstein

From Louise, Toronto, Canada

[1952/1953]

Dear Mr. Einstein,

My name is Louise, and I am ten years old. My Daddy thinks you are a very wonderful man. One of the greatest who ever lived, and so do I. Since you are such a great man, I thought people would like to have your picture with your autograph on it. I could have a raffle and sell tickets at 25 cents each. The winner would get your picture. Then I would send the money to the United Jewish Appeal. Would you please send me two pictures with your autograph so I could give one to my Daddy because he is a great man too.

Louise

P.S. My Daddy read the letter and he said the least I could do was to send you one dollar out of my bank for packaging. I am sending you one dollar, if this is not enough please write back and I will send you more.

Yours very truly,
Louise

P.P.S. Don't worry if your to busy to send me the pictures. Just give the dollar to any charity you like.

From Sandra, Elkins Park, Pennsylvania

March 18, 1953

Dear Dr. Einstein,

Our class is studying about the universe. I'm very much interested in space. I would like to thank you for all you have done so we can learn. I wish you a very happy birthday and many more.

Love,
Sandra

This letter and the ones below, among others, were sent for Einstein's seventy-fourth birthday.

From Marcia, Elkins Park, Pennsylvania

March 18, 1953

Dear Dr. Einstein,

I want to wish you a very happy birthday. The reason I am writing this letter is because I am interested in your job. Our class has your picture on our bulletin board. Our class is studying about the universe. I think studying about the universe is fun. I am glad you were born because if it wasn't for you we won't know as much as we do now. I am 9 years old.

Love,
Marcia

From Michael, Elkins Park, Pennsylvania

March 18, 1953

Dear Dr. Einstein,

I would like to wish you a very happy birthday. Our class is studying about the universe and I am very interested in it. I am very interested in your theory about light bending. We took a bottle of water and put a stick in it and saw how light bent it. If you ever have a little time (which I don't think you will) will you please send me some information? When I grow up I would like to be a scientist.

Sincerely,
Michael

From Lorraine, Los Angeles

June 4, 1953

Dear Einstein,

When I was a little 5 year old girl I saw your picture. I asked my mother who that nice man was. When she told me I asked her if I could send it to you because maybe you didn't have one. She said you probably had a lot of them so I kept the picture myself. When I was 7 I found out you were born under Pisces the same as me. I've always wanted to write to you.

Love,
Lorraine

P.S. I am now almost 9.

From Richard, New York

December 11, 1953

My Dear Prof. Einstein,

Today in school my class reported on famous immigrants. One committee had to report on your life and why you are famous. We got into a discussion of your theory. We found out that your theory is about relativity, but we don't know what relativity is.

I would appreciate it if you could send me some information best explaining relativity and what it is, and also some interesting facts about your life.

Thank you,
Richard, Age 11

Einstein wrote some fairly simple articles about relativity theory, but even these are hard for a nonscientist to understand, especially a child. A strong background in physics and mathematics and an understanding of the scientific words involved are necessary to comprehend the special and general theories.

Briefly, the special theory deals with the way things move but does not consider the pull of gravity on them. One of Einstein's observations has to do with the fact that when you talk about the motion of an object, you also need a frame of reference in relation to which the object moves; otherwise, you wouldn't know there is any movement at all. You've probably become aware of this yourself when you're in a plane and the land below is your frame of reference. The distance you're moving depends on the time that's elapsed while you're moving, so time and space (distance) are related. The special theory also deals with the conversion of mass into energy. A mass (the amount of matter something possesses) can be converted into a huge amount of energy that has been stored in the mass: for example, when a particle is released from an atom, it is converted to energy, as in a nuclear reaction that can cause a tremendous explosion. In the 1930s and 1940s, scientists worked on ways to convert the mass into energy, which finally led to the atom bomb.

The general theory of relativity is a generalization of the special theory and includes the effect of gravity on the shape of space and flow of time. Among other things, it says that matter, such as celestial objects, causes space to curve due to their gravitational pull. Therefore, for exam-

ple, a ray of light does not indefinitely travel in a straight line but bends in space and might eventually return to its point of origin.

These kinds of theories are used by scientists to work out complicated ideas about basic relationships in nature—of space, time, energy, mass, force, velocity, and so on. With the help of advanced mathematics, they can derive even more theories about relationships and interactions in nature and about the cosmology (origins) of the universe. Research in these areas has led to many practical scientific and technological innovations as well. Before Einstein, people did not realize that some of these entities, such as mass and energy, are related.

From Mary Ann, Carroll, Iowa

December 16th, 1953

Dear Prof. Einstein,

I am a littel girl 10 years old. My father is a doctor and my mother is a nurse.

I have heard that you like cookies. I heard a

story about you that makes me think you do. I don't know whether the story is true or not, but I'll tell you what I have heard about you liking cookies.

There was once a little boy who was having trouble in school with arithmetic. All of a sudden he greatly improved in his arithmetic. His teacher asked him why he had improved so quickly. The little boy said that it was because a good friend of his had helped him with his arithmetic. Also the little boy said that he gave his friend the cookies his mother baked and put in his lunchbox in payment for helping him with his arithmetic. He also said that his friend was very fond of the good cookies which his mother often baked. The teacher asked the little boy who his good friend was, and he told her that it was Prof. Einstein. I am in the fifth grade in school and I am having a lot of trouble with long division. I don't get along with arithmetic near as well as I do with painting, dancing, and playing the saxophone.

I think my mother makes cookies that are just as delicious as those which that little boy's mother made. So I asked her to make some cookies especially for you. I guess you are too far away to help me with long division, but I hope you will enjoy these Christmas cookies anyway.

Merry Christmas,
from Mary Ann

From Daniel, Buffalo, New York

January 28, 1954

Dear Professor Einstein,

In our sixth grade in School 66 we divided into different groups to study famous people who were born in Europe. A great many of us chose you. Our group knows how highly educated you are, so we don't know how to express these questions: Could you tell us about some of the things you do and teach at Princeton University? Could you write some of the easier parts of your theory in an easier form of English?, so we could read it? We like to read about your life. We think that you have contributed many great things to the world.

I know that all the children in the Einstein Study Group and I would appreciate it if you responded to our letter.

Sincerely,
Daniel

From a Boy in Berlin, Germany

April 16, 1954

Very esteemed Professor Einstein,

I am 10 years old. I read the *Berliner Morgenpost* and through this newspaper have learned something about your activities in Princeton. (When I heard the name "Princeton" I immediately looked for the town in the atlas and found it to be in the state of New Jersey.)

Furthermore, I heard that you are a very modest man despite being a professor. You have much great knowledge, and I am sure that you won't be angry with me if I ask you what your beliefs are about the origin of the world. I would also be very happy if I found an especially attractive postage stamp [on your letter]. That would give me great pleasure while also enlarging my stamp collection.

I wrote this letter all by myself.

Respectfully,
W. S.

The original letter was written in German.

From Clotilde, Rio Piedras, Puerto Rico

January 30, 1955

Dear Mr. Einstein,

My father is going to a meeting of the Ford Foundation at Princeton next Thursday. My brother, who is eleven years old, is going with him. His name is Jaime. We are Puerto Ricans at the University of Puerto Rico. My brother wants to see you but my father and mother say that he can only see you from a distance on your way to work, and that he should be respectful of your solitude, whatever else. The way they talk makes me think it must be an invasion of privacy.

I'm a girl twelve years old. When I was eleven, my father took me to Jamaica to see the Queen of England from a distance too, but here in this university they all say she's nice, but that you are more useful to humanity.

Very truly yours,
Clotilde

To the 5th Grade,
Farmingdale Elementary School,
Farmingdale, New York

March 26, 1955

Dear Children,

I thank you all for the birthday gift you kindly sent me and for your letter of congratulation. Your gift will be an appropriate suggestion to be a little more elegant in the future than hitherto. Because neckties and cuffs exist for me only as remote memories.

With kind wishes and regards,

Yours sincerely,
Albert Einstein

This was one of Einstein's last letters, written about three weeks before his death, and twelve days after his seventy-sixth birthday.

Afterword

I hope that this short book on Einstein's life, particularly as it relates to children and his own youth, will create more interest in this important scientific and historical figure among children and their parents and teachers.

It is evident from all of Einstein's writings that, besides his scientific work, most important to him was the idea of peace and harmony in the world. The writings show that scientists can bring not only creative ideas into the world, but exert a humanitarian—and very human—influence as well. Einstein thought of himself as an international figure, universal in thought, at home almost anywhere in the world, and interested in all of its peoples and cultures. He was an idealist in his view that a world government was the only solution to everlasting peace, but nevertheless valiantly spoke out in favor of it until he died. He detested snobbery and racism and managed to maintain his humility even when he became an international celebrity, happiest with the simplest of things such as a pipe, his small, plain sailboat, animals, and music. He worked endlessly for humanitarian causes that he hoped would end misery in the world, especially for his Jewish brethren as they tried to overcome the effects of the prejudices of the last century. He spoke out in favor of educational systems that stimulated children to be curious about their world, rewarded good teachers, and did away with the military-like discipline that prevented a free exchange of ideas between teachers and students.

Those who met Einstein never failed to mention his humility, the twinkle in his eye, and his good sense of humor and uproarious laughter. He had his faults, like all of us humans. Still, such a man, one who overcame many difficulties and setbacks in his life in the face of enormous burdens and unwanted publicity, has left a legacy matched by few and remains a role model for generations of young people to come.

Additional Reading

The following books are suitable for the general reader. Some are out of print but are often available used, over the Internet or in secondhand book stores.

Bernstein, Jeremy. *Einstein*. New York: Penguin, 1978.

Calaprice, Alice. *The Expanded Quotable Einstein*. Princeton, N.J.: Princeton University Press, 2000. (Original ed. pub. 1996.)

Clark, Ronald. *Einstein: The Life and Times*. New York: Crowell, 1971.

Dukas, Helen, and Banesh Hoffmann. *Albert Einstein, the Human Side.* Princeton, N.J.: Princeton University Press, 1979.

Einstein, Albert. *Cosmic Religion, with Other Opinions and Aphorisms.* New York: Covici-Friede, 1931.

———. *Ideas and Opinions.* Trans. Sonja Bargmann. New York: Crown, 1954.

———. *Albert Einstein/Mileva Marić: The Love Letters.* Trans. Shawn Smith. Ed. Jurgen Renn and Robert Schulmann. Princeton, N.J.: Princeton University Press, 1992.

———. *Einstein on Humanism.* New York: Carol Publishing, 1993.

Fölsing, Albrecht. *Albert Einstein.* Trans. Ewald Osers. New York: Viking, 1997.

Highfield, Roger, and Paul Carter. *The Private Lives of Albert Einstein.* London: Faber and Faber, 1993.

Hoffmann, Banesh. *Albert Einstein, Creator and Rebel.* New York: Viking, 1972.

Jammer, Max. *Einstein and Religion.* Princeton, N.J.: Princeton University Press, 1999.

Jerome, Fred. *The Einstein File: J. Edgar Hoover's Secret War against the World's Most Famous Scientist*. New York: St. Martin's, 2002.

Moszkowski, Alexander. *Conversations with Einstein.* Trans. Henry L. Brose. New York: Horizon Press, 1970.

Overbye, Dennis. *Einstein in Love: A Scientific Romance.* New York: Viking, 2000.

Paterniti, Michael. *Driving Mr. Albert.* New York: Random House, 2000.

Regis, Ed. *Who Got Einstein's Office?* Reading, Mass.: Addison-Wesley, 1987.

Rosenkranz, Ze'ev. *Albert through the Looking Glass: The Personal Papers of Albert Einstein.* Jerusalem: The Jewish National and University Library, 1998.

Sayen, Jamie. *Einstein in America.* New York: Crown, 1985.

Schwartz, Joe. *Einstein for Beginners.* New York: Pantheon Books, 1979.

Zackheim, Michele. *Einstein's Daughter: The Search for Lieserl.* New York: Riverhead Books, 1999.

For Young Readers

Hammontree, Marie. *Albert Einstein: Young Thinker.* New York: Simon and Schuster, 1984.

Lepscky, Paolo Cardoni. *Albert Einstein: Famous People.* New York: Barron's, 1992.

McPherson, Stephanie Sammartino. *Ordinary Genius: The Story of Albert Einstein.* Minneapolis, Minn.: Lerner Publishing, 1997.

Parker, Steve. *Albert Einstein and Relativity.* Broomall, Penn.: Chelsea House, 1995.

Index

Page numbers in *italics* refer to illustrations